Leifur Eiríksson
and Vínland the Good

Anna Yates

Iceland Review

LEIFUR EIRÍKSSON AND VÍNLAND THE GOOD
© Anna Yates   1993
Published in 1993 by Iceland Review ©
Reykjavík, Iceland.
Design: Magnús V. Pálsson
Litho: Prentmyndastofan, Reykjavík
Printed in The Netherlands

All rights reserved. No part of this publication may be reproduced, stored in a retrieval system, or transmitted, in any form or by any means, electronic, mechanical, photocopying, recording or otherwise, without the permission of the publisher.

ISBN 9979-51-072-2

Front cover:
Gaia, a replica of a 9th-century Viking Ship, sailed the North Atlantic in 1991.
Photograph by Páll Stefánsson.

Contents

1. From Iceland to the New World 9
2. Go West, Young Man! 17
3. Exploration and Violent Death 27
4. Face to Face with "Savages" 33
5. Where Was Vínland? 43
6. L'Anse aux Meadows: Proof at Last 55
7. Nordic Navigators 63
8. The End of the Greenland Settlement 75
9. How Much Did Columbus Know? 83
   Bibliographical Note 87

for Jim

Leifur Eiríksson finds America

Leifur was tossed about at sea for a long time, and he came to lands of whose existence he had not known. There were fields of self-sown wheat, and vines. There were also trees called maple. They took some of all these things with them, and some of the trees were so big that they were used for building houses.

Leifur found some men on a wrecked ship, and transported them all home and gave them hospitality all winter, showing what a great and kind man he was. He brought Christianity, and saved the people's lives. He was called Leifur the Lucky.

From the Saga of Eiríkur the Red

Author's Foreword

As the observant reader will already have noticed, this book has neither footnotes nor appendices, and hence makes no claim whatever to be a work of scholarship.
My aim has simply been to tell the story of this great adventure, for the general reader who may know little of the historical background. For those who read these pages and find they want to learn more, I recommend the books listed in *Further Reading* on page 88, most of which have footnotes in plenty.
The passages of the *Saga of Greenlanders* and the *Saga of Eiríkur the Red* which appear here are my own translations, from the collected edition of *Íslendingasögur* (Sagas of Icelanders) published by *Svart á hvítu* in 1987.
I have opted in favour of modern Icelandic spelling of personal names and placenames throughout. This involves the following three additional letters of the alphabet:
Æ,æ pronounced like *i* in *mine*
Ð,ð pronounced like *th* in *this*
Þ,þ pronounced like *th* in *path*
Accented and unaccented vowels need not concern the reader until we come to confront the question whether V*í*nland was really V*i*nland, in Chapter 5.
I would like to express my thanks to my publisher, Haraldur J. Hamar, for encouraging me to write this book, and to Don Brandt, Magnús V. Pálsson and Elín Jónsdóttir of Iceland Review.

I am most grateful to the following people for their assistance: Dr. Ólafur Halldórsson of the Árni Magnússon Manuscript Institute, who very kindly read the manuscript and made many helpful suggestions, Dr. Árný Sveinbjörnsdóttir of the University of Iceland Science Institute, who informed me about the climate of Greenland in the Viking Age, and Helgi Skúli Kjartansson, Cand. mag., who made constructive comments.

I thank my father, Sqdn. Ldr. Reg Yates, RAF (Retd) and Commander Tony Fanning, RN (Retd), who provided invaluable assistance by carrying out calculations of latitude and made useful contributions on matters navigational.

Anna Yates

Chapter One

From Iceland to the New World

Leifur Eiríksson: a name to conjure with. Whether called Leifr or Leif, Eiríksson, or Erikson, or simply Leif the Lucky, Leifur has come to symbolize the pioneering Viking spirit. Icelandic born, perhaps of half-Norwegian parentage, Leifur emigrated as a child to settle an unknown new land in the west, Greenland.
Yet when, as a young man, he heard of other lands in the west, Leifur was eager to sail still farther, to explore new oceans and new lands. He was probably the first European to set foot on American soil, to explore and establish a settlement on the mysterious continent which was to capture the European imagination in centuries to come.
The Nordic discovery and settlement of North America took place around the year AD 1000, nearly 500 years before Christopher Columbus found what he called the "Indies" far to the south. While Columbus has long been regarded as the discoverer of the New World, it is now acknowledged that Leifur Eiríksson discovered it first. Since 1964, Leifur Eiríksson has been commemorated each year in the USA on October 9, Leif Erikson Day, just as October 12 is marked as Columbus Day.
Two Icelandic sagas, the *Saga of Greenlanders* and the *Saga of Eiríkur the Red* (also known as the *Saga of Þorfinnur karlsefni)*, recount the main events in the settlement of Greenland and subsequently of "Vínland," yet neither can be counted as wholly reliable historical

sources. When the *Saga of Eiríkur the Red* describes an encounter with unipeds, for instance, we can safely assume we are entering the realms of fantasy.

All indications are that the *Saga of Greenlanders* is older and more authentic than the *Saga of Eiríkur the Red*. In all probability, the *Saga of Greenlanders* was written down around AD 1200, while the *Saga of Eiríkur*, many scholars believe, reworked and reinterpreted the theme somewhat later.

The *Saga of Greenlanders* has survived only in a single manuscript, interpolated into the *Saga of Ólafur Tryggvason* in the famed *Flateyjarbók* (Flatey Book), a priceless illuminated collection of sagas which was preserved for a time on Flatey island off the west coast of Iceland. In the 17th century, the book was presented to the Bishop of Iceland, who passed it on to the King of Denmark (at that time also King of Iceland). The Flatey Book then came into the keeping of the Royal Library in Copenhagen, where it remained until 1971, when Danish authorities returned the historic manuscript to Iceland. The Flatey Book now has pride of place at the Árni Magnússon Manuscript Institute in Reykjavík.

Two slightly different versions of the *Saga of Eiríkur* have survived; *Hauksbók* (Haukur's Book) and an untitled manuscript known by its classification number, AM557 Quarto, both of which belonged to the collection of Árni Magnússon (1663-1730). A professor at the University of Copenhagen, Árni Magnússon made it his life's work to save Iceland's medieval literary heritage from oblivion. One of the few Icelanders of his time who achieved a position of respect and influence among the Danes who then ruled Iceland, Árni Magnússon searched out old manuscripts and books on farms all over Iceland when he travelled around the country on official business in 1702-12.

To the poverty-stricken Icelanders of the time, centuries-old manuscripts had no great value. Had men of

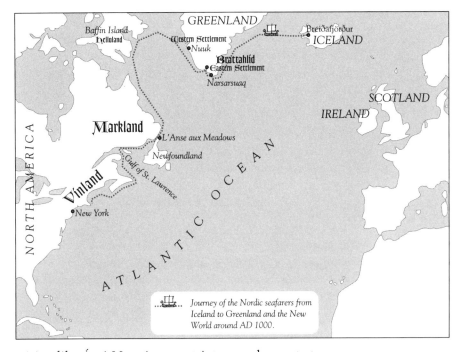

Journey of the Nordic seafarers from Iceland to Greenland and the New World around AD 1000.

vision like Árni Magnússon not intervened, a vast store of literary treasures might have been lost for ever. His collection had a narrow escape in 1728, when a great fire in Copenhagen consumed the bulk of his library. But most of the manuscripts were saved from the inferno, and when Árni Magnússon died he willed his entire collection to the University of Copenhagen. The original Árni Magnússon Manuscript Institute was established in Copenhagen in 1760. The Icelandic Manuscript Institute, founded in Iceland in 1962, changed its name to the Árni Magnússon Manuscript Institute in 1971, when the first of the precious manuscripts, preserved in Denmark for centuries, came "home" at last.

Hauksbók and the AM557 Quarto, which contain similar but not identical versions of the *Saga of Eiríkur*, are believed to derive from a common source, but probably not the original manuscript of the saga. *Hauksbók* was

11

The Saga of Eiríkur, preserved in Hauksbók, tells how Leifur was blown off course and found a land of vines and wheat.

penned in about 1302-10 for Haukur Erlendsson, a lawyer, and partly in his own hand. The saga was probably originally written down around the mid-13th century.

While the *Saga of Greenlanders* seems generally more reliable than the later version of the story in the *Saga of Eiríkur*, the relationship between them is not quite so simple. They tell stories which are the same in essentials, but which differ in almost every detail. The same events happen in a different order and context, and even to entirely different characters, in the two sagas. This has led to the tenable hypothesis that the sagas were composed independently of each other, but both based ultimately on the oral accounts of the discovery and settlement.

The *Saga of Greenlanders* focuses entirely on events in Greenland and on the expeditions to the western world. Eiríkur the Red, for instance, is simply "Eiríkur the Red of Brattahlíð." In the *Saga of Eiríkur* we are told his descent and that of his wife, how he came to leave Norway, where he lived in Iceland, and the details of the dispute which led to his outlawry. Some scholars, however, dispute the veracity of Eiríkur's Norwegian background, and favour the claim of *Íslendingabók* (The Book of Icelanders), which states that Eiríkur was an Icelander born and bred.

Guðríður Þorbjarnardóttir, heroine of the saga, appears unheralded as a shipwrecked waif in the *Saga of Greenlanders*, while the *Saga of Eiríkur* tells in detail of her descent, her father's pride and wealth, and how they came to follow Eiríkur to Greenland. In this respect, the *Saga of Eiríkur* adds a good deal to the bare bones of the narrative in the *Saga of Greenlanders*.

Perhaps there is a genealogical section missing from the beginning of the *Saga of Greenlanders* (of which only one manuscript has survived). Or this difference of emphasis may simply reflect growing interest in genealogies in the developing saga tradition. On the other hand, one author of the story may have been better informed about the Icelandic background of the characters than the other. He may even have set out, conscientiously, to research the background of the leading personalities of the Greenland-Vínland adventure. At least with respect to Guðríður Þorbjarnardóttir and her husband, the genealogical account should be reliable: Guðríður's descendants were eminent Icelanders, and relatively few generations had passed by the time the saga was written down. Icelanders, then as now, were well versed in family trees. For purely legal reasons, they had to be able to trace their descent back for five generations, i.e. from great-great-great-grandparents.

In Iceland itself, there has been a strong tradition of saga fundamentalism, whose essential creed was that every word of the sagas was true. The opposing view, that it was all fiction, has also been maintained with passion. The story of the Vínland discovery has been particularly hard-hit by the skeptics. Over the centuries, these sagas have often been dismissed as pure invention by scholars who could not accept that Vikings might have successfully conquered the perils of the Atlantic Ocean a thousand years ago. Only the discovery of an indisputably Viking-age settlement in Newfoundland in 1961 finally silenced the doubting Thomases.

Vikings have, in general, received a bad press: they have been portrayed as wild blood-thirsty vandals, plundering and pillaging along the coasts of Europe, carrying off slaves and ransacking churches. The men known as Vikings did their share of plundering and pillaging between the 8th and the 11th centuries, but the image of a bellicose warrior race is exaggerated. We must bear in mind that all contemporary accounts of the Vikings' savagery are drawn from chronicles written by monks and other clerics. The pagan Vikings tended, naturally enough, to raid churches and monasteries: after all, that was where the valuables were. So the conventional image of the terrifying and merciless men from the North is one-sided, to say the least.

Mainland Scandinavia was becoming overpopulated at this time, and economic pressure led Nordic farmers and seafarers to search for new lands to settle. As "Vikings" they roamed the seas, not only as pirates but also as merchants. They made their way east through Russia and the Baltic lands as far as the Byzantine Empire, where Viking warriors made up the emperor's élite Varangian Guard in Constantinople (now Istanbul). Many chose to put down roots in the northern regions of England, the Isle of Man or in Ireland, while others ventured farther afield, to the Faroe Islands and to Iceland, where they found uninhabited islands they could make their own.

The search for new land ultimately led the Nordic seafarers beyond the limits of the known world, to Greenland and the New World. Once they had found a place they could call their own, the Nordic "Vikings" transported their families, livestock and chattels to their new home. Afterwards, they could largely lay aside their swords in favour of ploughshares, although going a-viking remained a test of manhood and prestige.

In England, placename evidence strongly suggests that the Norse "invaders" tended to settle down in peaceful coexistence with the native population, laying claim to lands that had previously been uninhabited, rather than expelling the rightful owners with fire and the sword. In the Faroes, Iceland and later Greenland, the Nordic settlers set up their own form of republican government, which was to last until these outlying islands came under colonial rule from Norway in the 13th century.

First sighted by chance from a Greenland-bound ship, blown off course on the North Atlantic, Vínland was perceived as a paradise on earth by the Nordic settlers. But their attempt to live in the land of wild grapes and cornfields was short-lived. Hostile indigenous tribes, far outnumbering the settlers, made life too hazardous, and the would-be pioneers gave up and returned to Greenland and Iceland. The Greenland settlement, too, was to disappear, though not until the 15th century, and in somewhat mystifying circumstances.

While many details of the Nordic discovery of America remain shrouded in mystery, the sagas recount the essentially true story of Leifur Eiríksson and his brothers Þorsteinn and Þorvaldur, of Þorfinnur karlsefni and his wife Guðríður Þorbjarnardóttir, and of their son Snorri, the first European born in the New World. These brave and adventurous men and women, ever eager to journey into the unknown, opened up new horizons.

In 1930, when the Icelanders celebrated the millenium of their parliament, they received a gift from the people of the USA: a statue of Leifur Eiríksson, as visualised by Stirling Calder.

Chapter Two

Go West, Young Man!

Tradition says that Leifur Eiríksson, son of Eiríkur (or Erik) the Red, was born, probably in the 970s, on his father's farm, Eiríksstaðir in Haukadalur in the west of Iceland. According to the *Saga of Eiríkur the Red*, Eiríkur was Norwegian by origin, like many of those who founded the Icelandic nation. Between AD 870 and 930, Scandinavians in search of new lands, together with a fair number of Celtic slaves, made their way to this uninhabited green isle in the North Atlantic.

Eiríksstaðir in Haukadalur perpetuates the name of trouble-making Eiríkur the Red, who went on to found the Greenland colony.

Eiríkur and his father, says the saga, had fled their native country "because of some killings." (The *Book of Icelanders*, however, calls him simply "a Breiðafjörður man" and some scholars prefer this version.) Eiríkur seems to have been a difficult man to get along with: more killings made Haukadalur too hot to hold him, and he moved on to Breiðafjörður, where history repeated itself. Eventually, he was sentenced by the local *þing* or assembly to three years' outlawry.

Not in the least subdued by this experience, Eiríkur devoted his outlaw years to exploration. Many years before, an Iceland-bound seafarer named Gunnbjörn had glimpsed land to the west of Iceland, which had been known ever since as *Gunnbjarnarsker* (Gunnbjörn's Reefs). Eiríkur wanted to know more. He sailed far and wide along the coasts of the land in the west, and returned to Iceland after three years, determined to settle in this land which he called *Greenland*, "because he said that it would encourage people to go there if the land had a good name."

The following spring, Eiríkur and his family, presumably including the young Leifur, set off to Greenland, with several more shiploads of families with their household goods and livestock, seeking a new life in the west. Eiríkur made his home at Brattahlíð (near modern Narsarsuaq), where he started a new and respectable life as informal chieftain of the Nordic settlers in Greenland.

Once in Greenland, the hot-headed Eiríkur apparently settled down to the dignified life of a country gentleman. None of his three sons – Leifur, Þorvaldur and Þorsteinn – exhibited his aggressive personality, but his daughter, Freydís, seems to have inherited some of her father's less endearing traits.

Bjarni Herjólfsson became one of the Greenland settlers more by accident than design: the *Saga of Greenlanders* tells that he sailed home from Norway to Iceland, only to find to his dismay that his family had gone to Greenland with Eiríkur. Although it was late summer, and neither Bjarni nor any of his crew were familiar with the Greenland route, they decided to follow.

Once out at sea, they were caught in fog and northerly gales for many days. "After that they could see the sun, and could get their bearings," says the *Saga of Greenlanders*. After a day under sail, they sighted land: hilly and wooded, this could not possibly be Greenland. An-

other land farther north was spotted two days later. This land was low-lying and wooded, nothing like Greenland, with its great glaciers. After three days, they saw a third land, high and mountainous with a glacier, but Bjarni still did not feel that it resembled the descriptions he had heard of Greenland. Finally, they struck lucky, making landfall in Greenland after four days at sea.

Bjarni Herjólfsson's sighting of hitherto-unknown lands in the west led to enormous interest and speculation in Brattahlíð, although Bjarni came in for a certain amount of criticism for his lack of initiative in making no further exploration. Bjarni Herjólfsson, apparently, did not live up to the exacting standards of the adventurous Eiríkur and his band of sons. Once having seen America, he gave up the seafaring life, and settled down to a life of farming with his father. His reluctance to go ashore and explore is, however, quite understandable in view of the fact that the summer was almost at an end. He naturally wanted to reach safe harbour before sailing conditions

Eiríkur the Red and his fellow-colonists set sail for Greenland in about AD 985 from Breiðafjörður, on the west coast of Iceland.

deteriorated, especially since his ship must have been laden with goods from Norway.

Leifur Eiríksson, on the other hand, was itching to make new discoveries; he bought Bjarni's ship and gathered a crew of 35 to sail in search of these mysterious lands.

They found the lands without difficulty: first they came to the barren, rocky land which Leifur named *Helluland* (Stone Land). Next, they reached a flat, wooded land, with white sandy beaches, which Leifur called *Markland* (Wood Land). After another two days at sea, they went ashore on an island, where they tasted the dew on the grass, "and they felt they had never tasted anything sweeter." The sweetness of the dew hints at the riches to come. Having sailed through a strait and found safe anchorage for their ship in a lake just upriver from the shore, they made camp, and subsequently decided to spend the winter.

There was an abundance of salmon in the river, larger than the Icelanders were accustomed to. No frost came during the winter, and they concluded that they would be able to keep their livestock out at pasture throughout the winter. Clearly, they had travelled far to the south, and the *Saga of Greenlanders* confirms that there were more hours of daylight in midwinter than in Greenland or in Iceland: "The sun was visible at supper time and breakfast time at the winter solstice."

This description must have been much clearer to the medieval audience of the saga than it is to us. In the pre-chronometer age, the terms *eyktarstaður* (supper time) and *dagmálastaður* (breakfast time) were fairly accurate measures of time. Unfortunately, we can no longer be sure of their precise meaning, but this sentence probably means that there were seven or eight hours of daylight at the solstice. In southern Iceland, by contrast, the sun only rises above the horizon for about four hours on the shortest day of the year.

Once the explorers had built their houses at Leifur's

Camp, parties were sent out on day-long exploratory trips inland. One evening, a member of one of the parties did not return, and Leifur led a search party to look for him. When they found the missing man, a "southerner" named Tyrkir, he came towards them, babbling incomprehensibly in German. Eventually he started to make sense, in the Norse language, and told them he had found vines and grapes. Leifur was skeptical, but Tyrkir retorted, "Of course it's true! In my country there are plenty of vines and grapes."

Leifur and his men felled timber and gathered grapes to fill their ship, as well as the ship's boat, before sailing home to Greenland. As they departed, Leifur gave the new land a name to reflect its natural riches: *Vínland* (Wineland).

On the journey home, Leifur spotted a group of shipwrecked seafarers on a reef: fifteen Icelanders, bound for Greenland. Leifur took them aboard, and sailed home to Brattahlíð. After this serendipitous rescue on the high seas, he was always known as Leifur "the Lucky."

A good enough story, you may think. The story of Bjarni Herjólfsson's chance sighting of America is, however, contradicted in the alternative version of the Vínland story, recounted in the *Saga of Eiríkur the Red*. In this version, Leifur had been at the court of King Ólafur Tryggvason in Norway, and was bound for Greenland as a Christian missionary in the king's name, when he was carried off course. "He saw lands whose existence he had never dreamed of. There were fields of corn, and wild vines, and maple trees, some of them so large that they could be used for building houses."

The rescue of the shipwrecked travellers is also mentioned, but Leifur's soubriquet "the Lucky" is attributed not only to the rescue but also to his successful conversion of the Greenlanders to Christianity.

The high-minded young missionary Leifur, setting off to save souls in Greenland, makes an attractive image. It is this Leifur, clutching a crucifix to his breast, who is depicted in Stirling Calder's symbolic statue of the pioneer, situated on Skólavörðuholt in Reykjavík.

Sadly, this image must probably be dismissed. Scholars of Old Norse history and literature have reached the conclusion that the *Saga of Greenlanders* is the older, and in general more authentic, of the two sagas, probably written in about 1200.

The *Saga of Eiríkur the Red* is a more sophisticated and polished narrative, but the saga's claim that Leifur was a Christian missionary is decidedly spurious. Around the year 1200, strenuous efforts were being made to beatify the warrior-evangelist King Ólafur Tryggvason, who was popularly supposed to have converted several countries to Christianity – at the point of a sword: not only his own realm, Norway, but also Iceland, the Orkneys, Shetland and the Faroes. Not content with writing panegyrics on his real conversions, would-be historians were, by 1200, claiming that King Ólafur had also converted a sixth country, Greenland.

This re-writing of history led to the invention of an entirely new version of the discovery of Vínland, throwing out Bjarni and placing Leifur in the leading role. This makes, of course, for a much neater story, with one hero instead of two. But not a true story. It is probably safe to assume that the *Saga of Greenlanders* is closer to the truth, and that Bjarni Herjólfsson was the first to see America, while Leifur, the pioneer, was inspired by Bjarni's adventure to explore and settle in the New World.

Once the missionary Leifur is rejected, however, a whole new set of problems raises its head: the dating of the discovery of Vínland in AD 1000 is an old-established tradition, probably not unconnected with the fact that Iceland accepted Christianity in that momentous year. The date is, however, based solely on the putative connection of Leifur with King Ólafur, who died in AD 1000.

The sagas themselves give no clues whatever, and all that can be gleaned from *Íslendingabók* (the Book of Icelanders) is that Eiríkur left Iceland 14 or 15 years before the conversion of the Icelanders, i.e. about AD 985-6.

Failing other references, genealogical evidence can be used to give an indication of dating. While Eiríkur's and Leifur's descendants are forever lost in the mystery of Nordic Greenland, the descendants of Guðríður Þorbjarnardóttir and Þorfinnur karlsefni are well documented in Icelandic records. Tracing back the generations indicates that the young couple were probably in Vínland around AD 1030, in which case Leifur can hardly have been there before AD 1020. Since we can probably assume that he was a young, single man at the time of his expedition, he is unlikely to have been 35 or older, so he cannot have been born before AD 985. This leads to the inescapable conclusion that he was probably born *in Greenland*, after his father had emigrated from Iceland.

The question of Leifur Eiríksson's "nationality," though entirely anachronistic, has been fiercely disputed in the 20th century. He has been claimed as a "Norwegian" on the basis that his father was a Norwegian immigrant to Iceland (a doubtful claim in itself), while the tradition that he was born in Iceland means that he has generally been regarded as "Icelandic." What is absolutely certain is that neither term would have meant anything to Leifur, who would have called himself a Norseman and a Greenlander. And there are no Nordic Greenlanders left to stake their claim to Leifur Eiríksson and his achievement.

Bjarni Herjólfsson finds America

They set off out into the Greenland Sea, and sailed for three days until they lost sight of land. Then the favourable breeze dropped, giving way to northerly winds and fog. They did not know where they were for many days. After that, they saw the sun and got their bearings. They raised sail and sailed all that day before seeing land. They discussed what land it could be, but Bjarni said he did not believe it was Greenland. They asked whether he wished to sail to this land or not. "Let us sail in close to land." They did so, and soon saw that it was a land without mountains, wooded, and with low hills. They turned with land on the starboard side and with the sheet (corner of the sail) towards land.

They sailed for two days before seeing another land. They asked Bjarni whether he thought it was Greenland, this time. He said he did not think it was Greenland either, "for they say that there are great glaciers in Greenland."

They soon approached this land, and saw that it was flat and wooded. Then the wind dropped. The crew said they thought it would be a good idea to go ashore, but Bjarni was unwilling. They claimed they needed both wood and water. "You are not short of either," said Bjarni, but his crew were discontented.

He told them to raise the sail, and they did so. They turned the prow from land, and sailed on a southwesterly breeze for three days, before seeing the third land. It was high and mountainous, with a glacier. They asked if Bjarni wanted to go ashore, but he said he did not wish to, "for it looks like a land which has nothing to offer."

They did not haul down the sail, but sailed along the shore and saw that it was an island. They then turned the stern towards land and sailed out to sea on the same breeze as before (southwesterly). The wind started rising, and Bjarni told them to reef the sail, so as to safeguard both ship and rigging. Thus they sailed four days. Then they saw the fourth land. They asked Bjarni if he thought this was Greenland or not. Bjarni replied, "This is like the descriptions I have heard of Greenland, and here we will go ashore."

From the Saga of Greenlanders

Leifur Eiríksson's Expedition

First they found the land which Bjarni had seen last. They sailed inshore, dropped anchor, and went ashore by boat. They did not see anything growing. Above, there were great glaciers, and below it the land was like a solid slab of stone down to the sea. The land seemed to have nothing to offer. Leifur said, "Unlike Bjarni, we have set foot on this land. I name it Helluland (Stone Land)."
Then they returned to their ship. They sailed out to sea, found another land, sailed close to shore and dropped anchor, put out a boat and went ashore. This land was flat and wooded. There were white sands in many places they saw, and the land sloped gently down to the shore. Leifur said, "This land shall be named for its qualities, Markland (Wood Land)."
From there, they sailed out to sea on a northeasterly breeze. They were out at sea two days before they saw land. They sailed in to shore, and reached an island north of the mainland. They walked up onto the island and looked around in the good weather. There was dew on the grass, and they put their hands in the grass and to their mouths, and they felt they had never tasted anything so sweet.
[...] They decided to stay there for the winter, and built large

houses. There was no shortage of salmon in the river, and it was larger than any salmon they had seen. The resources of the land were such that it seemed unlikely that animals would need any winter foddering. There was no frost in the winter, and the grass hardly withered at all. The days were more equal in length than in Iceland or Greenland. The sun could be seen both at supper time and at breakfast time at the winter solstice.

From the Saga of Greenlanders

Tyrkir finds Wild Grapes

Before they had gone far, Tyrkir came towards them, and was received with joy. Leifur soon perceived that his foster-father was in an excellent mood. [...]he asked, "Why have you been so long, foster-father, and why did you leave the others?"
He talked for a long time in German, and rolled his eyes and grimaced. But they did not understand what he said. After a while he said in Norse: "I had not gone much farther than you. I have news for you. I found vines and grapes."
"Is that so, foster-father?" said Leifur.
"Of course it is true," he said. "I was born in a country where there is no shortage of vines and grapes."
The following day, Leifur said to his crew, "Split up into two teams. You must work alternate days on picking grapes or gathering vines and felling trees, to make a cargo for my ship." And this was done. It is said that the ship's boat was filled with grapes. A cargo for the ship was cut down.
In the spring, they made ready and sailed away, and Leifur gave the land a name according to its resources, Vínland (Wineland)."

From the Saga of Greenlanders

Chapter Three

Exploration and Violent Death

Leifur was never again to travel to Vínland, so far as the sagas say. His father, Eiríkur, had died, and it was Leifur's task as new head of the household to take over the estate at Brattahlíð. But his brothers, Þorvaldur and Þorsteinn, both proved eager to seek their fortunes in the land of vines and cornfields.

Þorvaldur, says the *Saga of Greenlanders*, felt that Vínland should be explored more thoroughly, at which Leifur offered his brother the use of his ship to sail to Vínland. Þorvaldur sailed with a crew of 30, and found Leifur's Camp, apparently without difficulty. They wintered there, living on the fish they caught, and in the spring Þorvaldur sent a boatload of men to explore to the west. When they returned in the autumn, they reported that the land was beautiful and well wooded, with forests growing down close to the sea. The beaches were of white sand. There were many islands, and shallows along the coast. On one of the islands, far to the west, they had discovered a wooden structure for storing grain, but no other signs of habitation by man or beast.

Þorvaldur spent his second summer in Vínland exploring to the north, where he was unfortunate enough to be swept ashore in a gale, breaking the keel of the ship. After a lengthy delay while the vessel was repaired, Þorvaldur and his men sailed to the east, where they saw a wooded headland between rivers. Þorvaldur was instantly captivated by the place. "This is a beautiful place. I would like to make my home here," he said.

Þorvaldur's wish was to come true, but not in the way he expected, as is the ironical way of the sagas. No sooner had he spoken the lyrical words than the Native Americans made their first, fateful appearance on the scene.

Spotting three curious hummocks on the shore, the Greenlanders went to investigate. They found three skin boats (canoes, presumably), with three men sleeping under each. They seized all but one of the men, who managed to escape in his canoe. They then slew all the captives. Before long, the local aboriginals, contemptuously termed *skrælingjar* (savages) by the Vikings, were back in force, paddling towards them in their canoes, armed with bows and arrows. The explorers took hastily to their ship, to fend off the attack as best they could. Finally, the Native Americans withdrew, but not before Þorvaldur had been mortally wounded by an arrow. His companions gave him a Christian burial before sailing back to Leifur's Camp. The following spring, they loaded up their ship with grapes and sailed for home.

When he heard their news, Þorsteinn, the third of Eiríkur's sons, decided to sail for Vínland, like his brothers before him: not for the sheer adventure, nor in hope of profit, but to bring Þorvaldur's body home.

By this time, Þorsteinn was married to a woman named Guðríður Þorbjarnardóttir. An Icelander of gentle birth, she had emigrated to Greenland with her father. The two sagas, as in so many other details, are not entirely consistent in their treatment of Guðríður: in the *Saga of Greenlanders* she first appears as the wife of a certain Þórir, shipwrecked on the ocean reef to be rescued by Leifur. Later, we are told that Guðríður was "a fine-looking woman, and wise, and she knew how to conduct herself among strangers."

In the *Saga of Eiríkur the Red*, on the other hand, Guðríður is introduced with full genealogical pedigree from Vífill, "who was of noble descent," and had come to Iceland with Auður the Deep-minded, one of the pioneers who first settled Iceland. Vífill was grandfather to Guðríður Þorbjarnardóttir, who grew up at Laugarbrekka on the Snæfellnes peninsula in the west of Iceland. She was also foster-daughter to a certain Ormur and his wife Halldís at nearby Arnarstapi, close friends of her father.

Guðríður, says the saga, "was the most beautiful of women, and showed her strength of character in all that she did." Her father, proud as well as wealthy, refused to marry her to a rich suitor who asked for her hand, on the grounds that he was of low birth, the son of a mere freed man.

Guðríður made the perilous ocean crossing to Greenland in company with thirty people including her father and her foster-parents. The ship was caught in bad weather, slowing their progress, and then the seafarers fell prey to a fever. Half of them, including Guðríður's foster-parents, died at sea. According to the *Saga of Eiríkur the Red*, a seeress named Þorbjörg was travelling from farm to farm in Greenland that winter. When she visited the farm where Guðríður and her father were staying, she enquired whether any of the women present could chant the *Varðlokur* or Weird Song, which would call up supernatural forces. Guðríður reluctantly admitted that she knew the verses, but refused to take any part in such heathen ritual, "for I am a Christian woman." She gave way gracefully in the end, however. "Guðríður chanted the verses so beautifully and well, that everyone present agreed that they had never heard them sung so wonderfully."

The prophetess responded by foretelling a glorious future for Guðríður. "You will marry well here in Greenland, but that will not last, for all your roads lead to Iceland. You will establish a great and good lineage, and from your descendants will shine a bright light."

While the episode with the seeress reinforces the image of the good and pious Guðríður (with a lovely singing voice, to boot), we need not assume it has any basis in reality. The prophecy serves to underline the fact that Guðríður did indeed go on to do great things and establish an illustrious line of bishops and scholars.

If we assume that the *Saga of Greenlanders* is more accurate, though less fulsome, Guðríður was already a married woman when she arrived in Greenland, but she soon lost her husband, Þórir, to the fever. Her second marriage was to Þorsteinn, son of Eiríkur the Red.

Guðríður appears to have been the first woman to sail for Vínland, when she set off with her bridegroom, Þorsteinn, and their crew of 25. But Þorsteinn was never to see the New World. The travellers were caught in bad weather, and spent the whole summer at the mercy of the elements out at sea, without making any progress. Just before winter came, they managed to make landfall in western Greenland, where they took refuge for the winter.

During the winter, the deadly fever struck again, claiming many lives including that of Þorsteinn Eiríksson. The *Saga of Greenlanders* recounts a rather chilling episode when Þorsteinn, after he was dead, sat up on his bier to address his widow. Þorsteinn foretold her marriage to an Icelander, journeys to Norway and Iceland, and a pilgrimage to Rome. He also saw her as the founder of a lineage of fine, beautiful, good and "sweet-smelling" descendants. Dreams, premonitions and predictions were an indispensable part of any saga, but this section of the saga, like the seeress's prophecy in the other version of the story, tells us that interesting events are in store for Guðríður.

A First Encounter with Native Americans

In the spring, Þorvaldur sent the ship's boat with a crew of several men to the west, to explore during the summer. They found the country beautiful and wooded. The woods grew close down to the sea, and the sands were white. There were many islands, and shallows. They found no trace of men or animals, except on one island far to the west, where they found a wooden structure for storing grain. They found no other man-made structures, and they left, returning to Leifur's Camp in the autumn.

The next summer, Þorvaldur set off to the east with his trading ship, and to the north of the land. They were caught in windy weather near a headland, and the ship was swept ashore and the keel broke from under the ship. They stayed there for a long time while they repaired the ship. Then Þorvaldur said to his crew, "Let us raise the keel here, and call this Kjalarnes (Keel Headland)," and they did so.

They sailed on to the east, and into the mouth of a fjord and to a headland nearby. It was all covered with trees. They anchored their ship in a haven, put out gangplanks and went ashore. Þorvaldur went up onto the headland with all his crew. He said, "This is a beautiful place. I would like to make my home here." They returned to the ship, and saw, on the sand inshore from the headland, three hummocks. They went there and saw three skin boats, with three men under each one. They divided their forces and seized all but one, who escaped in his canoe. They killed the other eight, then returned to the headland, where they could see into the fjord. They saw several mounds, which they thought must be villages.

Then they grew so sleepy that they could not keep awake, and they all fell asleep. Then they heard a shout, which woke them all:

"Awake, Þorvaldur, and all your crew, if you want to live. Go out to your ship, and all your men, and put out to sea, as quickly as you can."

Innumerable skin boats came out of the fjord, and attacked. Þorvaldur said, "Let us put out our battle-screen on the ship, and defend ourselves as best we can, but without attacking them." They did so, and the savages shot at them for a while, then fled as fast as they could. Þorvaldur asked his men if they had been wounded. They said they were not wounded.

"I have been wounded under the arm," he said. "An arrow flew between the gunnel and the shield. Here is the arrow, which will bring me to my death. Now I advise you to make ready to depart again as soon as possible, but you must take me over to the headland which I found so habitable. Perhaps I was right when I thought of staying a while. Bury me there, with a cross at my head and my feet, and call the place Krossanes (Cross Headland)."

From the Saga of Greenlanders

Chapter Four

Face to Face with "Savages"

Guðríður, widowed so soon after her marriage, returned to the home of her brother-in-law Leifur, Brattahlíð, where a new and eligible suitor soon came sailing over the horizon. Þorfinnur karlsefni was a wealthy Icelander, who had been in Norway before sailing to Greenland. They were married the following winter, and before long their sights were set to the west.
"People talked about sailing to Vínland as much as before, and Karlsefni's men and Guðríður were keen to go. The voyage was planned, and he took on crew, sixty men and five women. They made an agreement to divide equally among themselves all that they might acquire. They took a variety of livestock with them, because they intended to settle there if possible. Karlsefni asked Leifur for his camp in Vínland, but Leifur would only loan the houses, not give them," says the *Saga of Greenlanders*. Leifur had not entirely abandoned the idea of returning to Vínland himself, someday.
When the 65 (or 140, according to the *Saga of Eiríkur the Red*) seafarers arrived in Vínland, they were lucky enough to find a beached whale, which kept them well fed for some time. Timber was felled and stacked to dry during the winter, and the settlers fished and gathered grapes and other foods.
Not until the following spring did the *skrælingjar* appear. Temporarily frightened by the bellowing of the

settlers' bull (an exotic animal they had never seen before), the Native Americans came to Karlsefni's door, carrying bundles of sables and other furs, which they offered to barter for weapons.

Einar Jónsson (1974-1954), a pioneer of Icelandic sculptural art, cast Þorfinnur karlsefni in the heroic mould.

Sensibly enough, Karlsefni would not allow any weapons to be traded. Instead, the settlers offered the aboriginals milk in return for their valuable furs. "So the savages carried away their goods in their bellies, while their bundles and furs remained with Karlsefni and his companions," says the *Saga of Greenlanders* dismissively. In the *Saga of Eiríkur the Red*, the settlers are said to have bought the furs in exchange for strips of cloth, but the principle is the same: like subsequent colonists in the New World, the Vikings had no qualms about trading paltry gifts for valuables, when they were dealing with mere "savages."

Realising that the indigenous inhabitants could prove a danger, Karlsefni and his men built a strong defensive palisade around their camp. It was in this state of semi-siege that Guðríður Þorbjarnardóttir gave birth to a son, Snorri: so far as we know, the first European born in the New World. When the tribesmen returned later in the year, in force, to barter more goods, the Icelanders did not open their gates.

Food was taken out to the aboriginals, who flung their bundles over the fence in return. This encounter culminates in a bizarre episode: Guðríður sat in the doorway of her house, with her son in his cradle. A pale, brown-haired woman dressed in black appeared, and asked, "What is your name?" Hearing Guðríður's name, she replied, "My name is Guðríður," then disappeared with a loud noise, at the very moment when one of Karlsefni's servants killed one of the tribesmen, who had attempted to seize his weapon.

Þorfinnur Karlsefni concluded that the "savages" would be back soon, and this time on the warpath. Picking an advantageous spot between the sea and the forest, for the battle which he now realised was inevitable, he sent ten men as decoys to attract the aboriginals' attention while the rest lay in wait. In the subsequent fight, many of the tribesmen were killed.

The Native Americans had never known anything like the Icelanders' iron weapons. "There was one tall, strong man among the savages, and Karlsefni concluded he must be their leader. One of the savages had picked up an axe, looked at it for a while, then struck at one of his comrades, who fell down dead. Then the tall man took the axe, looked at it for a while, then threw it as far out to sea as he could. Then they fled into the woods, and that was the end of their dealings."

This rather moving incident, recounting the confrontation of the "noble savage" with a weapon beyond his comprehension, is reduced to absurdity as retold in the *Saga of Eiríkur*: "The savages found one of those who had fallen, and an axe lying by him. One of them picked up the axe and chopped at a tree, and they took turns at chopping, and they thought it a valuable thing, which cut well. Then one of them took the axe and chopped at a stone and the axe broke. He thought it was useless, since it could not cut stone, and discarded it."

In spite of the "victory," Karlsefni had no stomach for this life in Vínland, living under constant threat from the indigenous inhabitants. He declared, the following spring, in the typically laconic manner of the sagas, "that he did not wish to stay there any more, but intended to return to Greenland. They made preparations for their journey, and took with them a great quantity of vines, grapes and furs."

From Greenland, Þorfinnur karlsefni and his family

sailed to Norway, where they sold their valuable cargo, "and they were well received, by the noblest men in Norway." Just before they cast off to return home to Iceland, a man from Saxony asked Karlsefni to sell him his *húsasnotra* (an ornament of some kind), which was made of maple wood from Vínland. Karlsefni was loth to part with it, but when the man pressed him to accept half a mark (four ounces or 250 grammes) of gold, he agreed, like a good Viking businessman, to sell.

Þorfinnur karlsefni and Guðríður Þorbjarnardóttir settled down at Glaumbær in Skagafjörður, north Iceland. After Þorfinnur's death, Guðríður took over the running of the farm until her eldest son, Snorri, was old enough to manage the estate. In her youth, Guðríður had experienced enough adventure to last most people a lifetime: she had crossed oceans, seen her loved ones and companions fall prey to fevers, participated in founding two colonies, married and lost two (or perhaps three) husbands, and confronted hostile tribesmen. But Guðríður was not ready for retirement. In middle life, she set off again.

"Once Snorri was married, Guðríður went abroad, and walked south, then returned to her son Snorri's farm. In the meantime, he had had a church built at Glaumbær. Guðríður became an anchoress and a nun, and lived there for the rest of her life." The phrase to "walk south" denotes a pilgrimage to Rome, a fitting end to the adventurous life of Guðríður Þorbjarnardóttir, who must have been one of the most widely-travelled women of her time. Already middle-aged, she travelled across Europe *on foot* for the sake of her immortal soul, before settling down, at last, to spend her twilight years in prayer and contemplation.

Guðríður and Þorfinnur's descendants went on to earn fame and distinction, as foretold: two great-grandsons and a great-great-grandson became bishops. Brandur Sæmundsson (Bishop of Hólar in the northern diocese

Eiríkur the Red's home at Brattahlíð, near modern Narsarsuak, became the unofficial capital of the Greenland colony.

1162-1201) was a great-grandson of Snorri, while his predecessor in office Björn Gilsson (1147-62) was descended from Björn (or Þorbjörn), son of Guðríður and Þorfinnur. Þorlákur Runólfsson (Bishop of Skálholt in the southern diocese, 1118-33) was a grandson of Snorri Karlsefnisson. Thanks to the Icelanders' abiding interest in genealogy, many an Icelander today can trace his or her descent from the American pioneers of a thousand years ago, Þorfinnur karlsefni and Guðríður Þorbjarnardóttir.

The saga writer calls Guðríður a *skörungur* (a woman of substance), and she certainly earned the accolade. In the male-dominated world of saga literature, Guðríður Þorbjarnardóttir is one of a handful of strong, memorable women, such as Guðrún Ósvífursdóttir in *Laxdæla saga*, Hallgerður Longlegs in *Njáls saga*, and the settler Auður the Deep-minded.

For all that they founded a great dynasty, Guðríður, Þorfinnur and their son have left no traces at Glaumbær. Today the place is famed for its folk museum, located in

Vínland colonists Guðríður Þorbjarnardóttir and Þorfinnur karlsefni finally settled down at Glaumbær in northern Iceland.

the fine 19th-century farmhouse. The visitor to Glaumbær may find it strange to think that this remote northern farm was the home, in the 11th century, of a man born in America, and of his mother, who had travelled 3,000km to the New World, and 3,000km south to Rome.

The departure of Karlsefni and Guðríður marks, to all intents and purposes, the end of the Vínland story as we know it. A sad and bloody epilogue, however, recounts the story of Freydís Eiríksdóttir, daughter of Eiríkur the Red and sister of Leifur, Þorsteinn and Þorvaldur. Not to be outdone by her brothers, Freydís received Leifur's permission to use his camp, then set off for Greenland with two Icelandic brothers, Helgi and Finnbogi, and two ships' companies. Like Karlsefni, she took both men and women on her expedition.

Freydís, however, proved cunning and treacherous. She refused to allow Helgi and Finnbogi use of Leifur's Camp, and ultimately tricked her husband into killing the two brothers and all their company. When her companions refused to kill the five women, she grabbed an axe and slew them with her own hands. This left her as sole owner of all the goods accumulated by the expedition. "If we should reach Greenland safely," she said, "I shall kill any man who tells of these events. We shall say that they remained here after we left." But truth will out, and Leifur found out what his sister had done. "I shall not give my sister the punishment she deserves, but I foretell that her offspring shall not thrive," he declared.

There is no sign that any further attempts were made to settle in Vínland, but the Greenlanders seem to have continued to sail there for several centuries, presumably for the valuable furs, wood and grapes. In 1347, an Icelandic annal mentions a small vessel with a crew of 17, bound from Markland home to Greenland, which was carried off course in bad weather and found safe harbour on the Snæfellsnes peninsula. This, 350 years after Bjarni Herjólfsson's first sight of Vínland, is the last Icelandic record of the Nordic New World. By this time, both Iceland and Greenland were under Norwegian rule, and the Greenland settlement was in decline. The great days of the Nordic seafarers were over.

Þorfinnur and Guðríður in Vínland

After the first winter came summer, and then they became aware of the savages. A great crowd of men came out of the forest. The settlers' cattle were nearby, and a young bull started to bellow very loud. This alarmed the savages, who fled with their bundles, which consisted of grey furs and sables and all sorts of fur. They came to Karlsefni's house, and attempted to enter, but Karlsefni had the doors barred. Neither understood the other's language.

Then the savages put down their bundles, opened them, and offered the contents for sale. They were most interested in bartering for weapons, but Karlsefni forbade his men to part with their weapons. Karlsefni considered how to act, and asked the women to take milk out to them. When they saw the milk, they wanted that and nothing else. So at the end of the trading, the savages carried away their goods in their bellies, while Karlsefni and his companions kept their bundles and skins. And then they went away. Karlsefni now had a strong palisade built around his homestead, and they took refuge there. At that time, Karlsefni's wife Guðríður gave birth to a baby boy, whose name was Snorri.

Early in the winter, the savages came to meet them, far more numerous than before, and offering the same merchandise. Karlsefni said to the women: "Take out to them the foodstuffs which are most plentiful, but nothing else."

And when they saw the food, they flung their bundles over the palisade. Guðríður sat in the doorway by the cradle of her son, Snorri.

From the Saga of Greenlanders

Face to Face with Native Americans

Karlsefni and the others sailed to the mouth of the river, and called the place Hóp (lagoon). They found self-sown fields of wheat in the low-lying places, and vines in the hillier places. Every stream was full of fish. There were many animals of all kinds in the forest. They stayed half a month, and saw nothing alarming. They had their livestock with them.
Early one morning, when they looked around, they saw nine skin boats. Wooden poles were being waved on the boats, making a noise like flails.
Karlsefni said, "What can this mean?"
Snorri answered, "It may be a sign of peace. Let us take a white shield and hold it up."
They did so. The others rowed towards them, and found them amazing. They went ashore. They were small men, and hostile looking, with ugly hair on their heads. Their eyes were big, and they had broad faces across the cheekbones. They stayed a while, staring, then rowed away, south of the headland.
They had built their houses above the lake. They stayed for the winter. No snow came, and all their livestock stayed out at pasture through the winter.

From the Saga of Eiríkur the Red

Cloth for Furs

When spring came, they saw early one morning that a mass of canoes came around south of the headland. They put up their shields, and then they traded. The people mostly wanted red cloth, as well as swords and spears, but Karlsefni forbade this. They offered dark skins in return for the cloth, and took a hand's span of cloth for a skin, and tied it around their heads, and then went away for a while. When the cloth began to run short, they cut it smaller so it was no more than a finger's breadth, but the savages gave just as much for it, or more.

Then a bullock that Karlsefni had ran out of the woods and bellowed loudly. The savages were frightened and ran to their boats and rowed away to the south. They did not come back for three weeks.

From the Saga of Eiríkur the Red

Ambush!

"I believe the savages will be back for a third time, hostile and in force," said Karlsefni. "My plan is this. Ten men go out onto the headland over there and show themselves, and the rest of us go out into the woods." The place where the battle was to take place was between the sea and the forest. They did as Karslefni had said. The savages came to the place according to Karlsefni's plan. There was a battle and many savages fell. There was one tall, strong man among the savages, and Karlsefni concluded that he must be their leader. One of the savages had picked up an axe. He looked at it for a while, then struck at his companion, who instantly fell down dead. Then the tall man took the axe, looked at it for a while, and flung it as far out to sea as he could. Then they fled into the woods as fast as they could, and that was the end of their dealings. Karlsefni and the others stayed all that winter, but in the spring Karlsefni declared that he did not wish to remain, and intended to go home to Greenland.

From the Saga of Greenlanders

Chapter Five

Where Was Vínland?

Anyone who reads the sagas of the Vínland adventure will inevitably ask, "Where was Vínland?" A good question, but a difficult one to answer. Any attempt to pinpoint the location of Vínland usually leads to another question mark (or several).
One of the major problems lies in evaluating the evidence of the sagas, winnowing the wheat from the chaff. The sagas contrive to be impressively detailed in one respect, tantalisingly vague in another. Some factors must undoubtedly be attributed to invention, wishful thinking, exaggeration or simply misunderstanding. Others seem, to the casual reader at least, to constitute cast-iron evidence.
Over the years, scholars and amateur enthusiasts alike have come up with one ingenious theory after another, identifying diverse sites along the entire Atlantic seaboard of North America as "Vínland." No one place fits all the evidence in the sagas, but some are more likely than others. It all depends which piece of the jigsaw puzzle is discarded in constructing the theory.
In principle, the description of the sagas suggests a southerly location. Not only are we told of fields of self-sown wheat, and the wild grapes which give Vínland its name, but we also have the statement that "conditions were so favourable there that they believed that their livestock would not need winter fodder. There was no

frost in the winter, and the grass hardly withered. Hours of daylight were more equal there than in Iceland or Greenland, and the sun was visible at suppertime and breakfast time at the winter solstice."

For all this to be true, the Icelandic explorers must have travelled a long way south: scholars have generally suggested that the wild grapes and cereals point to a location in the New England region. Even there, however, the description of frostless winters is rather far-fetched. For winters that warm, we would have to look at least as far south as 40°N: Delaware or Maryland, perhaps. On the other hand, climatic change must also be taken into account. We know that Iceland and Greenland were considerably warmer in the Settlement Age – the 9th to 11th centuries – than they are today. Could the frostless winters of Vínland reflect a similar change farther south?

Should the frostless winters be rejected as hyperbole? Leifur Eiríksson's father, Eiríkur the Red, shamelessly bestowed the attractive name of "Greenland" on the land he settled, simply because it was good public relations. Perhaps Leifur followed his father's example, deliberately making his new colony sound like a paradise on earth.

Alternatively, the frostless winters could be an example of the saga-writer's wishful thinking, as he painstakingly recorded the tale on parchment at home in chilly Iceland. Even the *Saga of Greenlanders* was only written down approximately two centuries after the events it describes; and word-of-mouth evidence can be distorted a good deal over a far shorter period. And the *Saga of Eiríkur the Red* is far more prone to flights of fancy than the older saga.

The description of longer days at midwinter has the ring of truth: it sounds like an authentic record of an accurate observation. If frostless winters are easy to invent, hours of daylight are not. If only we knew the exact

Knowledge of the Vínland adventure survived as part of the Icelandic world-picture for many centuries. Sigurður Stefánsson's 16th-century map (with Iceland placed firmly at the centre of the world) depicts Markland, Helluland and the "Promontorium Vinlandia," as well as "Skrælingjaland" (Land of Savages). Across the ocean, we see the more prosaic Norway, Ireland and Britain, while to the north lies "Jötunheimar" (Land of Giants).

meaning of the old Nordic "breakfast time" and "supper time," we ought to have no trouble at all in pinpointing the precise latitude of Vínland!

Unfortunately, Leifur's observations are of almost no use because his terms of reference are archaic and, frustratingly, are practically meaningless to us. Some historians interpret "supper time" as 3.30 pm, others as 4 pm or even 5 pm. Most agree that "breakfast time" was probably around 9 am. Logically, however, the time from sunrise to noon and noon to sunset *must* be the same, so we could be thinking in terms of 9 am to 3 pm, 8 am to 4 pm, 7 am to 5 pm, etc.

The information that hours of daylight are "more equal than in Iceland or Greenland" provides something that may almost be hailed as a clue in this context. Brattahlíð, at 60°N, is four degrees farther south than southern Iceland (Reykjavík, for instance, at 64°N). This small difference in latitude, however, makes a noticeable difference to the length of the shortest day of the year: in southern Iceland, the sun only appears for four hours on the winter solstice, whereas down south in Brattahlíð, the shortest day is two hours longer.

If the Vínland explorers felt that the difference in the length of the day was worth mentioning, it must have been significantly longer than the four to six hours with which they were familiar in midwinter. An extra hour would hardly have impressed them, given that they drew no distinction between hours of daylight in Iceland and Greenland. An additional one-and-a-half or two hours is perhaps the logical minimum.

Many a theory has been built upon this enigmatic sentence, and no new hypothesis will be presented here. But it is interesting to test out the theory that Leifur's observation of the hours of daylight could lead us to Vínland.

If we calculate the latitude where the sun is visible for seven hours on the winter solstice, we arrive at 55°N, in Central Labrador, which is indisputably rather chilly for Vínland.

Eight hours of daylight brings us to 50°N, i.e. Newfoundland, the very place where a Viking-Age site was excavated in 1961, at l'Anse aux Meadows. Newfoundland, however, has no grapes or self-sown wheat, let alone frostless winters.

Nine hours of daylight at the winter solstice, and we are at 42°N, in the region of New York or Boston. Wild grapes and wheat, yes. Frostless winters, no.

Ten hours of daylight on the shortest day takes us down south to 31°N, in southern Georgia or northern Florida, where the winters are certainly frostless.

In other words, depending on how we interpret the saga's observations, Vínland could be almost anywhere on the eastern seaboard of North America!

We are left with the general, but imprecise, impression of a southerly location. Even the notion of a "place in the sun" has, however, been a question of fierce dispute.

Those famous grapes, which gave Vínland its name, have proved a difficult mouthful to swallow. Wild grapes do not grow farther north than New England, and many scholars have had difficulty in believing that the Nordic explorers could in fact have sailed this far south. If the possibility of their reaching these warm latitudes is rejected, the grapes have to be either an invention or a misunderstanding.

This argument against the grapes can be supported with reference to the sagas. It is clear, for instance, that the saga writer himself knows nothing whatever about grapes or wine. In the *Saga of Greenlanders* it is the "southerner" Tyrkir who finds the grapes: "He talked for a long time in German, and rolled his eyes and grimaced. But they did not understand what he said." (Although the saga tells us that Tyrkir talked in German, this may or may not be accurate. Perhaps "German" simply meant "some funny foreign language.") Then Tyrkir gets a grip on himself, and tells them, in Norse this time, that he has found grapes, adding for good measure that he knows all about grapes, because they are common in his own country. The inference is clear: babbling, grimacing Tyrkir has been eating grapes, and is drunk!

Only someone who knew of wine and grapes (*vínber* – literally wine-berries) solely by repute, not by experience, would make this error, assuming that grapes were alcoholic *on the vine*. Other references to grapes and vines reinforce the impression of ignorance: when Leifur has been told of the grapes, he tells his men to "gather grapes or cut vines, and fell trees, to make a cargo for my ship." In the spring, the settlers sail home with a shipload of timber, grapes and vines, and the modern reader cannot but ask: why would they cut the vines? and surely the grapes would have been rotten by the following spring?

These doubts need not mean that there were no grapes

in Vínland. After all, Tyrkir, who knew about grapes, said they were grapes. Or is Tyrkir an invention, inserted to add authenticity to an unlikely tale? The unconvincing treatment of the grape theme in the sagas may reflect only the saga-writer's lack of knowledge. The inspiring name of Vínland may have led him to think of intoxicating fruits on the vine. Not knowing what a vine looked like, he may have imagined grapes growing on trees whose timber would be useful, like the other timber the explorers felled. The oral account of the Vínland travellers may have left him in doubt as to whether the grapes were harvested by picking the fruits or by cutting the plants. This would explain the odd phrase "to gather grapes *or* cut vines." On the other hand, perhaps the vines *were* useful in some way, for instance in making rope. The fact that we cannot quite make sense of the story does not necessarily prove that it is nonsensical.

If the settlers did indeed find grapes, gather them and transport them home to Greenland, they must have had some way of preserving them, even though the saga writer does not mention this. They may have sun-dried them to make raisins – this would make sense, in view of the Nordic tradition of drying foods for the winter and for long sea journeys. A more imaginative view is that Tyrkir, the southerner who knew all about grapes, could have taught these Viking adventurers the gentle art of wine-making. One can almost imagine them, barefooted in the sun, treading the grapes under Tyrkir's direction, then making barrels and waiting patiently for the wine to mature. But one would expect the saga writer to comment on this interesting addition to Nordic expertise. Disappointingly, there is no sign that the discovery of Vínland led the Greenlanders to take up the civilised pleasures of wine-bibbing.

The unsatisfactory nature of the grape theme in the sagas has led some students to reject the grapes entirely, producing various interesting conjectures on one of the few apparently indisputable aspects of the story, the derivation of the name of *Vínland*.

If we throw the grapes overboard, we are no longer committed to a southerly location, opening up new possibilities for the placing of Vínland. A suggestion which has gathered a considerable following over recent decades is that the name is not *Vín*land but *Vin*land. That little accent over the *i* may seem insignificant, but it makes the world of difference in Icelandic, which is only a slightly altered version of Old Norse, the language spoken by Leifur, Þorfinnur, Guðríður and the rest.

Vín is a loan word in Old Norse and Icelandic, derived from the Latin *vinum* (=wine). It occurs in Old Norse as early as in *Hávamál* (The Words of the High One), a collection of gnomic poetry which certainly predates the Vínland discovery by a century or two. So Leifur and his company must have known the word *vín*.

Vin, on the other hand, is an old-established word in Old Norse. Etymologically related to the Anglo-Saxon *wine* and the Old High German *wini*, *vin* originally meant something in the nature of "pasture, meadow." Later the meaning developed into "farm," and ultimately the word died out altogether. In modern Icelandic, the old word has been retrieved and dusted off to express an exotic, if related, concept: "oasis."

The theory has been advanced that the land in the west was named *Vin*land for its grassy countryside. This name, preserved in oral tradition, ceased to make sense to those who heard it, because the word *vin* was no longer in common use, and the name of the land was thus amended to *Vínland*, giving rise to all the starry-eyed musings on grapes.

The evidence for this is, however, rather shaky. In all probability, *vin* was already obsolete by the Viking Age. It was much used in placenames in the period AD 100-600, but by the Viking Age, "pasture" was generally expressed by the words *eng* and *tún*, and no *vin* names occur in Iceland or Greenland. On balance, the most straightforward derivation, from *vín*, seems the more likely.

This kind of confusion between accented and unaccented vowels does not arise in modern Icelandic. Foreigners sometimes have trouble remembering where to put their accents, but Icelanders never do. Accented and unaccented vowels have entirely different sounds: *i* is pronounced as in *bit*, while *í* is pronounced like *ee* in *screen*. No confusion is possible.

The exact pronunciation of Old Norse is not known, but the accented and unaccented vowels were originally long and short forms of the same vowel sound. Add to this the fact that the Icelanders did not yet have a written language at the time of the settlement in the New World, and we are left to speculate: could they have confused *vín* and *vin*? And if they could, did they?

In an extreme version of the anti-grape hypothesis, the grapes are rejected entirely, largely on the grounds that Þórhallur, one of the members of Karlsefni's expedition in the *Saga of Eiríkur*, composes a verse protesting that he has not received the wine he was promised, but only spring water to drink. A suggestion has even been made that the "grapes" of the story could represent a specific pasture crop, vetch, some varieties of which produce intoxicating effects. This offers a neat explanation for Tyrkir's babbling and rolling of the eyes: Tyrkir was not drunk, he was hallucinating!

If we cannot depend on the grapes, or the frostless winters, and since we are not in a position to interpret the observations of latitude in the saga, we have little real evidence to help us find Vínland. If the name means "Pasture Land," the field is wide open. Some scholars have, however, attempted to locate Vínland on the basis of the sagas' descriptions of sailing routes and topography, rather than by means of climatic clues.

It is a well-known fact that historical sources tend to

record only the out-of-the-ordinary, while taking the familiar for granted. This tendency gives history a built-in distortion factor, which means that we often know about kings, wars and disasters, while we rarely know what the ordinary peasant ate, or wore, or thought.

By the same token, the sagas leave unsaid what we most want to know about how the Vikings found Vínland. Methods of navigation must have been so obvious that they could be taken for granted by a contemporary audience. To the modern reader, however, the sagas' descriptions of the ocean journeys are little short of mystifying. According to the *Saga of Greenlanders*, Bjarni Herjólfsson sails from the west coast of Iceland, sailing for three days "until land was out of sight." A long interval of "many days" follows, during which the seafarers are caught in fog and northerly winds.

Once the weather clears and they can get their bearings, they sail for *one* day before reaching the first land (Vínland). With land on the port side and the *skaut* (corner of the sail) towards land, they sail *two* days before reaching the second land (Markland). They sail out to sea on a southwesterly wind for *three* days before sighting the third land (Helluland). This land is observed to be an island. They then turn the ship out to sea once more "on the same breeze" (still southwesterly, presumably), and sail *four* days in stormy weather before seeing the fourth land (Greenland, finally).

The formalized pattern of one day, then two, three and four to reach the first, second, third and fourth lands, is highly suspicious, to say the least. A taste for literary symmetry seems to be indicated here. Whatever the value of the account, we are at least told of southwesterly winds, and we also have general descriptions of the appearance of the three new lands in the west. When adventurous Leifur Eiríksson sets off to retrace Bjarni's ocean journey, we read only that "they prepared their ship and sailed out to sea, and they came first to the

51

land which Bjarni had found last." Bjarni *must* have given Leifur fairly accurate sailing directions to the lands in the west, but the saga writer gives us no hint of what they were.

The descriptions of the three lands and their geographical features may, however, provide some clue to locations. Helluland is totally barren, with bare rock between the glaciers and the sea: the obvious conclusion is that this describes Baffin Island: more specifically, the Cumberland Peninsula, which lies almost directly opposite the Western Settlement.

Markland is low-lying, wooded country, sloping gently down to the sea, with white sands in many places on the shore. The pale-coloured sand would, of course, be memorable to men born in Iceland, where the sands of the shore are almost invariably black. Markland is often identified as Labrador, but this is a controversial issue; the location of Markland depends, naturally enough, on how far south we are looking for Vínland. If we believe the clues which point to a southerly location for Vínland, Newfoundland would be a tempting identification for Markland.

From Markland, the voyagers sail on a northeasterly breeze for two days before they sight land once more. They land on an off-shore island, north of the mainland, where they taste the dew and admire the view. After this they sail into the strait which divides the island from a headland which extends to the north from the mainland. Low tide leaves their ship high and dry in the strait, but they are eager to go ashore. A river flows down into the strait from a nearby lake, and at high tide they return to the ship to move it upriver, and anchor the vessel on the lake. This is the place which will be known as Leifur's Camp, where Leifur, his brother Þorvaldur, and Þorfinnur karlsefni all reside in turn during their sojourns in Vínland.

But the settlers do not simply sit at Leifur's Camp, waiting for something to happen. Leifur sends teams inland to explore, and Þorvaldur in his turn sends groups of men by ship to "explore to the west during the summer." Returning in the autumn, they report that the land is beautiful and wooded, with woods growing down close to the shore, and white sands. The sea is shallow, and there are many islands. This expedition also finds the first signs of human habitation, a wooden store for grain, on an island far to the west. Although the explorers must often have stopped to go ashore and look around, a whole summer's travelling would surely take them a long way south of Vínland proper (wherever that was). If, as the sagas say, the journey from Greenland to Vínland via Helluland and Markland took only a matter of days, they could have covered immense distances during the months of summer.

Þorvaldur leads the second exploratory foray in person, heading north and east. When they run ashore in a storm, they break the keel of the ship, and name the place where they stay while carrying out repairs *Kjalarnes* (Keel Headland). Then they sail on to the east, into the mouth of a "fjord," and land on a wooded headland. Here they kill a number of Native Americans, and are rewarded by being attacked by hordes of vengeful aboriginals. Þorvaldur is killed in the fray, and is buried on the lovely headland, on which he bestows the name *Krossanes* (Cross Headland) with his dying breath.

In spite of the apparent topographical detail of these accounts, there are many places along the Atlantic coast of North America which fit (or can be made to fit) the bill. More or less persuasive claims have been made for Hudson Bay, Labrador, Newfoundland, the St. Lawrence estuary, New Brunswick, Nova Scotia, New England, Massachusetts, Rhode Island, Long Island Sound, Virginia, Georgia and even Florida.

The truth is that we simply cannot tell what to accept and what to discard when we read the sagas, and on the basis of the sagas alone we cannot say with any confidence, "*this* was Vínland." If we take the relatively obvious course of regarding the grapes, wild grain and mild climate as believable, New England is a reasonably good fit. But in view of the circumstantial descriptions of islands, peninsulas, straits and headlands, there is something to be said for the St. Lawrence estuary, which abounds in promising natural features.

The only hope for a reliable identification of Vínland lies in archaeology. Archaeologists have already proved the crucial point that Nordic people in fact reached North America in the Viking Age, but the site in question, l'Anse aux Meadows in Newfoundland, cannot be Vínland, no matter however creatively the evidence is interpreted.

The search is not over, though. Perhaps, at some time in the future, archaeologists will unearth a Nordic site by a shallow lake near the sea, with a rune-stone bearing the legend "Leifur was here." Stranger things have happened.

Chapter Six

L'Anse aux Meadows: Proof at Last

For many centuries, the sagas recounting the discovery of America by Nordic seafarers around AD 1000 were generally regarded as very dubious sources. Various eminent men, most famously the Arctic explorer Fridtjof Nansen, declared that the whole story was no more than a myth.
Skeptics claimed to discern links between the sagas and the 9th-century *Voyage of St. Brendan*, a highly stylised account of a 6th-century Irish saint's spiritual odyssey across the ocean to an earthly paradise, the Land of Promise of the Saints. Like Vínland, this wondrous land abounds in grapes. It was assumed that the tale of St. Brendan (enormously popular reading in medieval Europe) had spawned a similar legend in Iceland.
Today, when archaeologists have proved that Nordic settlers did indeed reach North America in Viking times, even the Brendan story seems less far-fetched. Stripped of some of its religious symbolism, it seems to describe a journey via the Faroe Islands and Iceland, across an ocean strewn with icebergs, to the New World. In 1977, Tim Severin proved that medieval Irish sailor-monks *could* have made the journey, when he successfully sailed a replica leather curragh across the Atlantic.
As we have seen in the preceding chapter, nothing is easier than to demolish the evidence of the sagas. Not only do the *Saga of Greenlanders* and the *Saga of Eiríkur*

the Red differ on many crucial points; they are also vague, self-contradictory and occasionally incomprehensible.

Nonetheless, the essentials of the story are clear, and common to both versions: a Nordic settler in Greenland (Bjarni or Leifur) was blown off course when sailing from Iceland to Greenland. Finding an unknown land far to the south and west, he sailed northwards and eastwards, sighting two more lands, until he found his way to Greenland. The journey was retraced by one or more groups of aspiring settlers (Leifur, Þorvaldur), ending with an attempted settlement by Þorfinnur karlsefni and Guðríður Þorbjarnardóttir.

The problem faced by those who believed that the sagas could be trusted, albeit selectively, was that there was not a scrap of concrete proof that the Vikings had ever set foot in North America, 500 years before Columbus. The search for some kind of authentic relic of the Viking presence has been on since the last century, not least because members of the large Scandinavian-American communities felt an understandable desire to see *"their"* candidate vindicated in the nationalistically-inspired contest with Columbus.

This chauvinistic fervour produced some bizarre "evidence," ranging in nature from wishful thinking to downright fraud. The strange nine-metre-high Newport Tower in Rhode Island has long been claimed as a Viking structure, although there is absolutely no evidence to support this notion. While the tower seems to serve no good purpose, there is no reason to suppose it a Viking folly rather than an English one.

Enormous public interest was aroused when a massive stone slab carved with runic letters was "discovered" in 1898 by a farmer of Swedish origin in Minnesota. It seemed to support revolutionary theories that the Vikings had travelled far into North America, rather than having kept to the eastern seaboard.

Once submitted to expert appraisal, though, the Kensington Stone proved to be no more than an amateurish fraud, presumably perpetrated by the "discoverer" of the stone; though written in runic letters, the text was a potpourri of modern Swedish, Norwegian and English. Although entirely discredited, the Kensington Stone has given rise to a thriving tourist trade based on the supposed Viking presence in the Midwest.

Traces of Viking-Age weapons were "found" in 1930 near Beardmore in Ontario, but no proper excavation was made at the time, and some years later allegations were made that the objects had been planted. There is no doubt that they were authentic Viking relics, but several witnesses claimed to have seen them before the so-called "find," and stated that they had been brought to Canada from Scandinavia. Although fraud was never proved, the Beardmore Find remains highly questionable evidence.

Probably most famous of the Vínland hoaxes is the so-called Vínland Map, which turned up in 1957 in myste-

L'Anse aux Meadows proved to be an authentic Viking site, although nothing like the Vínland of the sagas.

rious circumstances. Supposed to date from the 15th century, the map shows Europe, Asia and Vínland, and states that Vínland was discovered by Leifur Eiríksson and Bjarni Herjólfsson.

The lack of a known provenance for the map naturally led to skepticism, although the parchment on which it was drawn was certainly old enough. The fact that the Vikings did not, so far as we know, draw or use maps also gave rise to doubts, while the most important stumbling block to acceptance of the map was that Greenland was shown, with suspicious accuracy, as an island. There is no evidence that the Nordic settlers ever travelled far enough north to discover this fact. Controversy raged for many years, but analysis of the ink finally proved, to most people's satisfaction at least, that the map was a 20th-century forgery.

Following all these misguided attempts to vindicate Leifur Eiríksson's claim to fame as the discoverer of America, the uncovering of a genuine, indisputably authentic, Viking-Age Nordic site in Newfoundland in 1961 was all the more electrifying.

Mounds indicating the remnants of ancient buildings at l'Anse aux Meadows, at the northern tip of Newfoundland, were locally known as "Indian Camp." The place was brought to the attention of Helge Ingstad and his wife, archaeologist Anne Stine Ingstad, when they were scouring the Atlantic coast of North America for possible Nordic settlement sites.

During the seven summers from 1961 to 1968, the Ingstads uncovered the remains of eight buildings of Nordic type, with walls of stacked turf like medieval Icelandic and Greenlandic buildings. An international team of archaeologists from Norway, Sweden, Canada, the USA and Iceland (including Dr. Kristján Eldjárn, who went on to become President of Iceland) came together to dig in Newfoundland.

Artefacts found in the buildings, such as a ring-headed

bronze pin, a glass bead, a soapstone spindle-whorl, a quartzite needle-hone and a bone needle, all led to the conclusion that this was a Viking-Age site.

The excavation of a smithy, however, was what clinched the identification of l'Anse aux Meadows as a Nordic site. The residents had smelted iron from bog-iron, plentiful in the vicinity. And it is known that the Native Americans had no knowledge of iron or smelting until after the European settlement. The sagas themselves tell that the "savages" had never seen anything like the iron weapons wielded by the Nordic intruders. Here, finally, was solid proof that the Nordic mariners had been in the New World.

None of the buildings at l'Anse aux Meadows was for livestock, which seems to exclude the possibility that this was a conventional long-term settlement. Land-seeking Icelanders and Greenlanders always took livestock with them. The presence of women's tools like the spindle-whorl and the needle-hone indicate, however, that the community included women, and a small wooden arrow may mean that there were children.

Excavations yielded a large number of iron rivets, which had clearly been cut through, plus many tiny fragments of wood, preserved in the nearby bog. Archaeologists deduce that ship repairs must have been carried out at l'Anse aux Meadows. Old rivets are sometimes found where fragments of ships' planking have been burned for fuel, but then they are not neatly sliced apart. Only a small quantity of iron was ever smelted at l'Anse aux Meadows, though, and this is one of many clues which lead to the conclusion that the site was not inhabited for long.

Middens, those invaluable rubbish-dumps of history, are small, and, even more important, there is no trace of a cemetery. The Nordic settlers, certainly Christian by the time of the New World settlement, would have wanted respectable burial for their dead. If l'Anse aux

Meadows was only inhabited for a matter of years, the problem may not have arisen; alternatively, the grieving relatives may have opted to transport bodies home to Greenland for a pious Christian burial by a priest.

Relics such as charcoal fragments, which had been used as fuel in the smelting process, were subjected to radio-carbon testing, which yielded dates of AD 700-1000. These dates reflect, not the precise date of the fire, but the age of the wood, which may vary considerably. Trunks of mature trees will give an earlier date than twigs and saplings. The date range points to the Viking Age, and this is corroborated by other indications, such as the artefacts uncovered and the ground-plan of the buildings. These, which consist of three clusters of buildings (plus the smithy, which stood apart), are of an early type of turf building.

Further excavations were carried out at l'Anse aux Meadows in 1973-76, and it has been classified as a National Historic Park since 1977. Administered by Parks Canada, today l'Anse aux Meadows draws thousands of visitors, to see the replicas of three Viking-Age buildings and re-live this once-forgotten chapter in the history of North America.

The l'Anse aux Meadows excavations, extraordinary though they were, produced no direct link with Leifur Eiríksson or any other named Nordic settler. Whether l'Anse aux Meadows was a permanent homestead or a short-term staging-post, it bears no resemblance whatever to the Vínland of the sagas.

No wild grapes or cereals grow this far north, there is no natural harbour, and access from the open sea is far from easy, although the sea level has dropped considerably since Viking times, and the settlement must originally have been closer to the shore. Studies of pollens from the site show that the area has not undergone any major climatic deterioration, so that grapes could not have thrived a thousand years ago any more than they can

Excavations at L'Anse aux Meadows revealed saga-age buildings and artifacts, plus evidence of iron-smelting.

today. No serious claim can be made for l'Anse aux Meadows as Leifur's Camp or any other named location in Vínland.

Two butter-nuts, found on the Nordic site, lend support to the idea that this was a staging-post en route to Vínland. These nuts are only found in a more southerly climate – no farther north, in fact, than the mouth of the St. Lawrence. So the residents at l'Anse aux Meadows clearly travelled on south, at least once. The traces of ironworking and repairs also fit in with the idea of a staging-post. To the far-ranging Nordic mariners, keeping their vessels in good repair was of the utmost importance.

The Ingstads' discoveries at l'Anse aux Meadows prove nothing, directly, about Leifur Eiríksson, Guðríður Þorbjarnardóttir and the rest. But they do prove that Nordic seafarers reached North America at the right period, and stayed there, at least for a while. Essentially, then, the sagas are true, however much the details may have been distorted in the telling and retelling.

If we believe in Vínland, the settlement in Newfoundland must mark a staging-post on the route to the wonderland in the south. If we reject all the mythical trappings of the sagas, perhaps Newfoundland was the real Vínland, before romantic speculation and exaggeration created the image of a wondrous land of wine and grapes. Either way, nobody can ever again dismiss the *Saga of Greenlanders* and the *Saga of Eiríkur the Red* as unalloyed fantasy.

Chapter Seven

Nordic Navigators

Skepticism was, for centuries, the rule rather than the exception when it came to Nordic claims to have discovered America around AD 1000. After all, Columbus had enough trouble crossing the Atlantic in 1492; surely he knew more than the "primitive, bloodthirsty" Vikings of the Dark Ages who raided the coasts of northern Europe raping and pillaging as they went.

Post-Renaissance man, with his preconceptions about the "Dark Ages," had difficulty with the concept of people who were warriors *and* visionaries, bandits *and* explorers. He could only perceive the Vikings as the destructive savages portrayed in many contemporary chronicles.

Yet the other side of Nordic culture is clearly recorded in the rich Icelandic literary heritage, the elegant poetry of the skalds, the splendours of Norse mythology. In Iceland, the freedom-seeking Nordic settlers of the 9th century created their own republican form of government, rejecting the rule of kings and princes. Valiant efforts were made to uphold a peaceful order, and discourage the rough justice of the sword.

The "Vikings" were a more complex phenomenon than has often been assumed. Strictly speaking, the term "Vikings" applies only to the Nordic seafarers who roamed the seas of Northern Europe from the 8th to the

11th centuries: essentially, they were looking for land to settle. They were only temporarily "Vikings" before settling down to become farmers.

Scandinavian seafarers created what may loosely be described as a Nordic empire, stretching south to France, and across the North Atlantic. Their extraordinary success can largely be attributed to two factors. In their longships, they had the best ocean-going vessels of their time, which meant that they could go farther, and faster, than their contemporaries. They also had unique navigational skills. At a time when other seafarers dared not venture out of sight of land, the Vikings knew where they were going, and could find their way back.

The Viking ships used to be something of an enigma; although they are often described in glowing terms in saga literature, and portrayed on innumerable stone carvings, practical detail was lacking. Not until several well-preserved Viking ships were discovered in the late 19th and early 20th centuries was it possible to make an accurate assessment of the seaworthiness of the famous longships.

Light was finally shed on the structure of the Vikings' vessels following two spectacular finds of grave ships in Norway, the Gokstad ship (excavated in 1880) and the Oseberg ship (excavated in 1904). Both were the richly-furnished graves of chieftains: a man of about 50 at Gokstad, a young queen (with an attendant) at Oseberg. While both ships were crushed and broken, the timber

Carvings of Viking ships, like the 8th-century Hunninge Stone from Gotland, gave only the sketchiest idea of their structure.

had been remarkably well preserved in both cases by blue clay. In the case of the Oseberg ship, even the high prow and stern had survived. After painstaking work to preserve the fragments which had survived a thousand years in the ground, it was finally possible to piece together two authentic Viking-Age vessels. A vast hoard of grave goods, equally well preserved, cast new light on many aspects of everyday life in the Viking Age.

Although both are "longships," the Gokstad vessel is clearly a sturdy ocean-going vessel, while the elegant, decorative Oseberg ship is not built for long journeys on the open sea, and must have been a pleasure-craft for travelling along the coast in fine weather.

Reconstruction of the ships (now on display at the Viking Ship Museum in Oslo) brought to light some of the special qualities of these unique vessels, which combine elegance with strength, lightness and flexibility.

Dating from the late 9th century, the Gokstad ship is 23.3 metres long, and 6.25 metres across its widest point. Amidships, it measures 1.95 metres from the bottom of the keel to the gunnel, and its draught is under a metre. A vast oak log has been used to form the massive keel, which strengthens the flexible hull and makes the ship more stable. Thin oak planking is built up from the keel to form a clinker-built hull (i.e. each strake overlaps the one below); the joins between the strakes are caulked with tarred cord. While each strake is riveted to the one above and below, strakes are lashed rather than riveted to the keel and to the ribs. This means that the hull remains supple, and can flex freely in the water.

The hull of the Gokstad ship is made up of 16 strakes, the first nine of which are especially thin at 2.6 cm. The tenth, which is at the waterline, is nearly twice as thick. It is riveted to the ribs, adding strength to the hull. The next three strakes are 2.6 cm thick, followed by a thicker one (3.2 cm) for the oar-holes. The top two strakes, which make this a particularly high-sided ship, are especially thin, 1.6 cm. The mast is a sturdy pine, 30 cm in diameter. The top had rotted away, so the height of the mast is not known with any certainty. Sail and rigging have not survived, so there is still plenty of room for speculation on exactly how the Viking ship was sailed.

On each side of the ship are 16 oars; narrow-bladed, they are made of pine, and vary in length from 5.30 to 5.85 metres. Each oar-hole has a shutter, so it can be closed in heavy seas. From the position of the oar-holes, about 40 cm above the deck boards, scholars have drawn the conclusion that the crew probably rowed sitting rather than standing. There are, however, no thwarts or other seats on the ship, so probably the Vikings sat on their sea-chests to row. On the Gokstad ship, the deck boards are simply laid on the cross-beams and ribs, allowing easy access to the space below for storage, baling, etc.

Older than the Gokstad ship, the Oseberg ship was probably built around AD 800. It is a little shorter, 21.44 metres, and narrower, 5.1 metres. The ship is far less sturdily built than the Gokstad vessel: the keel is joined together from two pieces, and the hull is low-sided, with only two strakes above the waterline. The oar-holes cannot be closed. The deck boards are also nailed down except for a few, fore and aft and on each side of the mast. This seems to indicate that there was little need for storage space on this vessel. The Oseberg ship has fifteen pairs of oars. Unlike the Gokstad ship, the Oseberg ship survived complete with its beautifully-carved prow and stern, which arch up to five metres above the waterline. The style of the carvings indicates that the vessel was built in about AD 800, although it was clearly old and dilapidated when committed to the earth as the final resting-place of a young queen.

Following painstaking reconstruction work, the Viking-Age vessels cast considerable light on the secrets of the Vikings' success. Their vessels were light, swift and manoeuvrable, whether under sail or oar power. The

The Oseberg ship lay a thousand years in the earth, before being excavated to shed light on the Vikings' sailing skills.

supple hull was strong without rigidity, so it could flex freely under the pressure of the waves, while the shallow draught meant that they could sail in to the shore, make a lightning raid of plunder and pillage, then make a fast getaway.

The question of how fast the ships could sail, and how well they could be handled on the high seas, could only be answered by carrying out sea trials. Only a few years after the discovery of the Gokstad ship, the first replica of a Viking ship set sail for America.

An approximate re-creation of the Gokstad ship, named the *Viking*, was sailed from Norway to Newfoundland in 1893. The journey took under a month, and the ship's captain, Magnus Andersen, recorded speeds of ten to twelve knots. The side-mounted rudder or steering-oar, which had seemed strange and awkward to the discoverers of the Viking vessels, proved itself highly practical and efficient.

Many more tests of the sailing qualities of Viking ships have been made in the past hundred years, all going to prove that the Nordic seafarers' ships were perfect vessels for their size. Their methods of construction cannot be applied to larger vessels, which must inevitably be built stronger, more rigid and therefore heavier. The massive wooden warships and merchant vessels of a later age have nothing in common with the spare and elegant efficiency of the Viking ship.

More Viking-Age ships have been recovered and restored in the past century, including several which were deliberately sunk in about AD 1000 to block the harbour at Skuldelev by Roskilde fjord in Denmark. Among them was a *knörr*, the Vikings' freighter. The Skuldelev *knörr* is shorter than a longship, and proportionately broader, 16.6 metres by 4.6. Like the longship, the *knörr* is clinker-built, with a shallow draught, but it has a capacious open hold amidships where livestock and goods can be stored. The cargo capacity of the *knörr* has been estimated at 15 to 20 tons. Naturally enough, since it is constructed for carrying loads rather than for speed or battle, the *knörr* is a stubbier structure than a longship, without the airy elegance of the Gokstad and Oseberg ships. The sailing qualities of the *knörr* were proved when a replica, *Saga Siglar*, was sailed around the world in 1984.

In all probability, it is the *knörr* which we must imagine making the long haul to Greenland and Vínland, when the settlers set off with their livestock, their household goods, their wives, children and, presumably, slaves. Unlike the longship, the *knörr* was not intended to be rowed over long distances; since the ship carried heavy cargo which occupied most of the space amidships, deck room for rowers was limited. The Skuldelev ship has only four oars.

As the millennium of the Nordic discovery of America approached, a joint Norwegian-Icelandic venture was launched in 1991 to commemorate the anniversary. The *Vínland Revisited* project, the brainchild of Norwegian businessman Knut Kloster, involved replicas of three different Viking ships, the Gokstad and Oseberg longships, and the Skuldelev *knörr*.

All three ships sailed together up the Norwegian coast, before the Gokstad replica, captained by Norwegian adventurer Ragnar Thorseth, set out on the transatlantic journey via the Orkneys, the Faroes, Iceland and Greenland to Vínland. Unlike the original Viking seafarers, the crew of the Gokstad replica (named *Gaia* for the ancient Greek earth goddess) were accompanied across the open seas by a more modern vessel, the *Havella*, equipped with all contemporary navigational aids.

On the North American coast, the *Gaia* was rejoined by the other two ships, which had hitched a lift by freighter across the Atlantic. They then sailed together down the American coast, to l'Anse aux Meadows, New York and up the Potomac to Washington DC in time for Leif Ericson Day on October 9, 1991.

Archaeology has told us much about the Viking ships, but the question of sails and rigging remains uncertain. Essentially, those who have created and sailed the replicas have deduced what sort of sail and rigging is probable. The ships, as we know from many stone carvings, had a single square-rigged sail. They are often depicted with some kind of ropes, which the crew are holding onto.

It has sometimes been assumed that the square sail was

a clumsy and cumbersome piece of equipment, which confined the Vikings to sailing with a following wind. This would, naturally, have limited their ability to sail where and when they wished. While the Viking ship's rigging does not have the quick-change manoeuvrability of a modern sailing yacht, which can tack in seconds, the Nordic seafarers could sail quite effectively with a sidewind. The ships manoeuvre well at up to 45 degrees to the wind, with the aid of a *beitiás* or tacking boom, which extends the bottom of the sail, stretching it to catch the wind.

The 1893 Gokstad replica skimmed over the waves at an average of 10 knots, and achieved up to 12 knots; the *Saga Siglar*, a *knörr*, averaged eight knots, and could do up to 12 in favourable conditions. These speeds, achieved under test conditions, demonstrate that the information given in the sagas about the length of various ocean passages is generally quite believable. While the perilous Atlantic crossings to the Faroes, Iceland, Greenland and Vínland might be lengthy in bad weather, they could be made in a matter of days in good conditions.

The formalized description of the four stages of Bjarni's journey (one day, then two, then three and then four) is a little too neat to be true. But the information that it took only ten days to sail from Vínland to Greenland is not necessarily suspect. The high-speed travel of the Viking seafarers leads us to speculate on exactly how far south they actually went. Some scholars have assumed that all suggested sites for Vínland which are south of New England are "too far" south for the Vikings to have reached them. But at the speeds we are considering, the Vikings could easily have sailed south to Florida, or even farther!

Whatever the truth of the matter, and however far the Vikings really sailed, their ships were not much good to them without navigational skills. As to how they found

An Icelandic-Norwegian crew sailed a replica of the Gokstad ship, the Gaia, across the Atlantic in 1991.

their way, the sagas are tantalisingly unhelpful. When the saga tells that Þorfinnur Karlsefni and his party "set off out to sea in their ship and soon reached Leifur's Camp safe and sound," one asks, in exasperation, *How?* Unlike Columbus and later explorers who crossed the Atlantic farther south, the Vikings would never have set out to cross this vast expanse of open sea at one fell swoop. Their method, a natural consequence of centuries of coastal sailing along the Scandinavian fjords, was to island-hop. They used no maps or charts, but it seems that experienced sailors could commit to memory, and then describe, quite complex routes across the ocean. Their "sea-craft," acquired through years of experience, enabled them to read signs which mean little to the modern mind. They observed changes in the colour of the sea, indications of fish and other marine life, clouds over distant (and invisible) landmasses, birdlife, etc.

A clue to this form of sea-craft is given by a passage of

Landnámabók (The Book of Settlements) written down in Iceland in the 12th century: "Wise men say that from Staður in Norway it is seven days' sailing to the west to Horn on the east coast of Iceland, and from Snæfellsnes four days' sailing to Hvarf in Greenland. From Hernar in Norway one should sail directly west to Hvarf in Greenland; one sails north of Shetland, so that it can only be seen in clear weather, and south of the Faroes, so that the upper half of the mountains can be seen over the horizon, then south of Iceland, where there are birds and whales. From Reykjanes in south Iceland, it is three days' sailing to Jölduhlaup in Ireland in the south; and from Langanes in north Iceland it is five days to Jölduhlaup in Ireland, and four days' sailing north to Svalbarði."

It is clear, even from the unhelpful sagas, that the Viking seafarers knew the world was round and could navigate to some extent by the sun. When the travellers lose sight of the sun, they are at a loss. Bjarni Herjólfsson, for instance, is caught in northerly gales and fog. But when the weather clears, "they saw the sun and got their bearings." The stars were less important to the Nordic navigators, as their journeys were invariably made in summer, when nights are light in northern latitudes and starry heavens are not to be seen.

Seafarers, it seems, could navigate along what we think of as lines of latitude: they knew how high the sun ought to be at midday at the relevant time of year in some place they knew (Iceland, for instance). From there, they could sail directly to east or west by reference to the sun. The account of Bjarni Herjólfsson's adventures supports this idea. When Bjarni has "got his bearings," he sails west until he runs into a landmass which is clearly not Greenland. He knows from the sun that he is much too far to the south, so he heads northwards. Thus, by trial and error, he finds Greenland.

The question of whether the Vikings had any kind of

navigational technology remains unanswered. There is no unequivocal proof that they had instruments for judging the angle of the sun. But in 1948, at an archaeological dig at Unartoq, Greenland, an interesting object turned up, on which many theories have been built. It is a half-disc of oak, about seven centimetres across, marked off at equal intervals around the circumference: were the disc complete, it would be marked off into 32 equal sections. Scholars have interpreted this as part of a simple bearing-dial. With a pin through the middle, it could be used to read off the angle of the sun, and thus take a bearing. On the other hand, this would only have been useful in conjunction with detailed knowledge of the changing angle of the sun at different latitudes and different seasons, and we have no proof that the Nordic seafarers had any such knowledge.

By the 12th century, observations of exactly this kind were being made in Iceland by the semi-legendary astronomer Oddi Helgason, known as Star-Oddi. *Odda tala* (Oddi's Table) records Oddi's observations and calculations of various phenomena, such as the changing

The Gaia arrived in Washington DC on Leif Eriksson Day in 1991.

angle of the sun at midday throughout the year, and the changing direction of dawn and sunset at different times of the year. Oddi's Table would certainly have been a useful aid to Atlantic navigators, and no doubt we can assume that Oddi would not have made the observations unless there was a need for them. The question remains whether some kind of systematic observation of this kind existed a century or two before Star-Oddi, at the time of the Viking voyages.

Navigational tables and the bearing-dial remain controversial, but neither is essential to the hypothesis that the Vikings had navigational skills. They could find their way quite adequately without instruments of any kind. They knew the sea and its signs, and if they could see the sun, they could steer a course to east or west. Familiarity with approximately where the sun *ought* to be would enable them to tell whether they were a long way north or south of their expected position.

As to how far west or east they travelled, they cannot have known the distances with any precision. Distances sailed are always measured in *dægur* in the sagas. This term may mean days or half-days; it must have reflected distances sailed in average conditions, and given experienced sailors an approximate frame of reference.

With supreme confidence in their ships and their own expertise, the Viking mariners took their lives in their hands as they set off unhesitatingly to face the perils of unknown oceans, convinced that the whole world lay at their feet. By guess and by God (or more probably Óðinn), they found their way.

Chapter Eight

The End of the Greenland Settlement

The Nordic discovery of the New World is indissolubly bound up with the history of the Nordic colony in Greenland. Only from Greenland was the North American mainland accessible.
On the face of it, the colonisation of Greenland by Icelandic settlers seems all but incomprehensible. One wonders why they would choose that cold and forbidding land, when the life they knew in Iceland must have been harsh enough.
But both Iceland and Greenland have seen dramatic climatic change since the settlement days. We cannot perceive either land with the eyes of those first arrivals, but recent research on the ice of the Greenland glacier has given new insight into the climatic fluctuations which proved so decisive for the Greenland settlement. When Greenland was settled, its climate was far milder than it is today, but this warm period was succeeded by a prolonged cold period when the Greenland climate was less hospitable than it is today.
When the first Nordic seafarers made their way to Iceland and attempted to settle, in the mid-9th century, average temperatures were low. Those first settlers, Raven-Flóki, Garðar Svavarsson and others, gave up in face of the harsh climate, and it was at this time that Iceland received its chilling name. But within a few decades, the climate had improved considerably, and

when the Icelandic settlement began in earnest around AD 875, with the arrival of Ingólfur Arnarson, the North Atlantic region was entering an unusually warm period. Iceland was, according to the sagas, wooded "from the mountains to the sea" when the settlers arrived, and it is estimated that about a quarter of the country, 25,000 km^2, was covered by woodland and coppice at the time of the Settlement. Today, by contrast, there are only 1,250 km^2 of woodlands in Iceland. In regions of Iceland which are today quite uninhabitable, settlers built their homesteads, and raised their sheep, cattle and crops.

When Eiríkur set off to explore Greenland in the late 10th century, the warm period was at its zenith. The new land in the west may not quite have deserved the name it was given by Eiríkur the Red "because it would encourage people to go there if the land had a good name." But it was no frozen wasteland. Southern Greenland had plenty of green pasture for the Icelanders' livestock, and it must have seemed much like home to them.

For independent-minded inhabitants of the young Icelandic commonwealth, the prospect of yet another virgin land, waiting to be settled, was tempting. A severe famine, which had struck Iceland a few years earlier, must also have increased the attractions of a "green" land in the west.

Eiríkur the Red was, as we have seen, a hot-headed misfit, who had trouble adapting to the relatively liberal demands of the newly-founded republic. When he was outlawed from Iceland, he simply went exploring to the west, and travelled widely along the Greenland coast, learning enough to be convinced that it was a place worth settling. He met no native residents, no Inuit ("Eskimos"), and he assumed that Greenland was uninhabited, as Iceland had been.

While the sagas record the names of only a handful of the settlers who followed Eiríkur to Greenland, we know that there were up to 300 farms, about 200 in the Eastern Settlement (around modern Narsarsuaq) and 100 in the more northerly Western Settlement (around modern Nuuk). We can therefore deduce that the Nordic Greenlanders must have numbered around 3,000. By comparison, the Icelandic population in 1100 was around 80,000.

Although they lived by farming, much as they had done in Iceland, the Nordic Greenlanders were also dependent on trade with Europe. Greenland was a treasure-house of valuables, coveted by kings and noblemen. Not only did Greenland provide furs, walrus ivory and narwhal tusks (probably the origin of all the medieval mythology surrounding unicorns), it was also a source of fine hunting falcons, and, the ultimate status symbol of the time, polar bears. Even the skin of a polar bear was priceless, while a live polar bear was a gift fit for a king.

The brief *Tale of Auðunn*, one of the short stories of the Saga Age known as *þættir*, tells of an Icelander who travels to Greenland, where he spends all his worldly goods to buy a polar bear. After various trials and tribulations, he reaches King Sveinn of Denmark, in order to present him with the bear. He spends some time at the royal court, after which the king offers him a position of respect as a steward. Auðunn refuses the honour, saying that he must return home to Iceland to help his aged mother. When he departs, the king rewards him for the gift of the bear by presenting him with a new-built ship, together with a bag of silver pieces, plus a gold ring from the king's own hand.

All over medieval Europe, kings and princes vied to acquire the rarest of treasures: a live polar bear from Greenland.

In the early years of the Greenland settlement, communications were frequent between Iceland and the west-

After 1408, when a splendid wedding was held in Hvalsey Church, nothing is known of the fate of the Nordic Greenlanders. But the church still stands.

ern outpost. Though the ocean journey might be perilous, it could be made in a few days in favourable conditions, and it became all but routine. Greenlanders sailed east to Iceland, on to Norway to sell their goods, and returned with commodities needed by the colonists. With good vessels and navigational skills, the North Atlantic was no obstacle.

It was at this time that the Greenlanders travelled even farther westward, when Bjarni Herjólfsson (or possibly Leifur Eiríksson) caught a chance glimpse of land in the west. The wooded lands to the west held promise for the Greenland settlers, who had no timber for building other than driftwood. Whether they found grapes or pasture (*vínber* or *vin*), they soon gave up on their attempt to settle the New World. Against thousands of Native Americans, a few dozen Europeans (even armed with invincible iron weapons) had no chance of victory. They had not come to conquer, but to settle. Their idea was not to take the land by force.

But journeys from Greenland to the New World seem to

have continued after the age of the explorers, and as late as the 14th century a small vessel, en route from "Markland" to Greenland, was washed ashore in western Iceland. The timber would have been worth the risk of the journey.

In the days of Eiríkur the Red and his sons, the future of Greenland must have looked rosy. A pleasant enough climate, rich lands to the west, endless resources of luxury goods to be sold in Europe in return for necessities like iron, and reliable sea communications with Iceland and Scandinavia. Greenland was important enough to be converted to Christianity (the legend says by Leifur, but probably by some nameless proselytizer). It had its own bishop, whose episcopal seat was at Garðar (modern Igaliko). Even today, traces can be seen of the vast stone cathedral built there in the 12th century.

Yet within two centuries of the foundation of the Greenland republic, things started to go wrong. Probably the most important factor was climatic change. The Icelandic settlers were unfortunate enough to have discovered Greenland at the peak of a remarkably warm period, when it provided the living conditions they were looking for. As soon as the Greenland settlement was established, temperatures started to drop, and the climate steadily worsened until around AD 1200, when average temperatures were fully 2°C lower than they had been in AD 1000. The Greenlanders were simply not equipped to cope with the environment in which they found themselves.

Cooler temperatures brought drift-ice farther south, all but cutting off the vital sea route to Iceland.

Weather conditions grew more severe, so agriculture and animal husbandry became more difficult. At the same time, the worsening climate made the sea journey to Iceland far more hazardous, as drift ice floated ever southwards; communications with Iceland became less dependable, and the Greenlanders grew more isolated in their increasingly hostile environment. As the ice moved south, the absent Inuit returned to southern Greenland, to compete with the Nordic Greenlanders for the resources of the land.

In 1261, Greenland came under the rule of the Norwegian crown, as Iceland did in 1262. The sea route to Greenland became a royal monopoly, and the life-supporting Iceland-Greenland trade route ceased to exist. Both Icelanders and Greenlanders were entirely dependent on the whims of their new masters in Norway.

But demand for goods from Greenland was in decline: elephant ivory (finer than tusks from Greenland) was becoming available from the Orient, as well as plentiful furs, fabrics and spices. The long, perilous journey to Greenland grew less profitable, and less tempting, than in the heyday of the polar-bear trade. Norway, too, was in trouble. In a trade war with the Hanseatic merchants, Norway was losing its supremacy at sea. The Black Death also took a heavy toll when it swept through Norway in 1349, hitting Bergen, the centre of the Greenland trade, particularly hard.

A single vessel, the "Greenland knörr," sailed to Greenland at irregular intervals, but when it sank in 1369 no new ship was commissioned. By this time, the Norwegians had all but lost their shipbuilding skills.

While Greenland travel was irregular at this period, journeys were made, both by accident and on purpose. In 1406, a Norwegian ship headed for Iceland was carried off course and landed in Greenland. The castaways did not get away until four years later.

News of the Greenlanders was unclear: rumours spread that they had "gone native" and renounced Christianity in face of the increasing difficulty of their way of life.

The smaller, more northerly, Western Settlement was abandoned in the 14th century, at a period when temperatures dropped to an all-time low.
Yet the Nordic Greenlanders seem to have clung on to their European way of life, unable to adapt and live off the land as the Inuit did. And the Greenlanders kept in touch with Europe, at least sporadically. By a fascinating chance, the cemetery at Herjólfsnes (modern Ikigait) preserved its dead and their clothing in permafrost; when the graveyard was excavated in 1921, it was revealed that the Nordic Greenlanders of the 15th century were buried, not in out-of-date Viking-Age attire, but in contemporary European fashion. Many wore the fancy liripipe hoods typical of the 14th century, while others were buried in caps and gowns with design details dating from the late 15th century. The clothing was made of homespun Greenland cloth, but the Greenlanders clearly kept up with developments on the fashion scene back in Europe.
By 1500, just as the new colonisation of the Americas was beginning farther south, the Nordic colony in Greenland had ceased to exist. The disappearance of the Nordic Greenlanders is something of an enigma, which has led to a variety of hypotheses. The climate had improved again by the late 15th century, though it was still cooler than at the time of the Greenland settlement. Perhaps the rising temperatures came too late, after centuries of harsh winters which had destroyed grassland and killed off livestock. We know that the Greenlanders were in a poor state long before the end: malnourished, suffering from the severity of the climate, and growing ever more isolated from other Scandinavian countries.
There were no Nordic Greenlanders left in 1500. But how did they disappear? Did an unusually harsh winter carry off the last debilitated survivors? Weakened by their difficult life, did they fall an easy prey to the Black Death or other disease? Or were they captured or slaugh-

tered by the pirates who are known to have raided Greenland, as well as many other places in the North Atlantic?

According to one theory, they were ultimately massacred by the Inuit, who were moving south with the advancing ice to reclaim their ancestral territory. The converse theory suggests that the beleaguered colonists eventually gave up the struggle to live like Europeans without any support from Europe, and threw in their lot with the Inuit.

It has even been postulated that the Nordic Greenlanders travelled west to the North American continent where they mingled with the Native Americans; this theory, though unsupported by any evidence, has proved tempting to some of the American apologists of the "Leifur lobby," who have been drawn to the idea of a continuous Nordic presence after the discovery of the New World.

Perhaps the last few Nordic Greenlanders made one final, despairing attempt to flee "home" to Iceland in face of hostile Inuit and the harsh environment. Since they no longer had the fine Viking longships at their disposal, they would have set off to dodge the ice floes in small vessels quite unsuited for ocean sailing. Their chances of survival would have been small.

We shall never know exactly how the Nordic Greenlanders met their end, but the gradual decline during the preceding three centuries makes their extinction an entirely natural end to the story. When the Vikings knew the North Atlantic like the backs of their hands, Nordic Greenland was a viable entity, and by no means isolated from the European mainstream. But under the pressure of both climatic and economic forces, the Nordic Greenlanders, on the furthermost periphery of the Nordic world, simply faded away. The rest is silence.

Chapter Nine

How Much Did Columbus Know?

After centuries of uncertainty, the Nordic discovery of America is today an established historical fact, although we will never know all the details of that discovery. And now we face an intriguing question: was there a link between the first discovery of America in about AD 1000 and its rediscovery in 1492?
The conventional version of history perceives Cristoforo Colombo of Genoa as a visionary, a man ahead of his time, who realised that the world was round while his benighted contemporaries still believed they could sail off the edge into the void. This is, of course, an over-simplification. The roundness of the world was no secret, even in the 15th century, although notions of the globe's size were far from accurate – hence Columbus's conviction that he was sailing to the Indies, rather than discovering a whole new continent. And Columbus was not merely following a dream. He had travelled widely during the years before the *Nina*, the *Pinta* and the *Santa Maria* sailed for the New World. He may even have come to know of Leifur Eiríksson's journey nearly 500 years earlier. Interestingly, Columbus himself claimed, in a biogra-

Legend claims that Columbus visited Iceland in 1477.

phy, *Historie di Cristoforo Colombo*, published posthumously by his son, that he had travelled to Iceland in February 1477, on a vessel sailing from Bristol in England. Local tradition holds that Columbus spent that winter on the farm of Ingjaldshóll on the Snæfellnes peninsula – in the far west of Iceland, and close to the very places where Eiríkur the Red, Guðríður Þorbjarnardóttir and the other westward-bound adventurers had set sail. Legends of "Vínland the Good" had never entirely died out in Iceland, and who knows what hints Columbus may have picked up during his sojourn in the north? The Iceland journey may be a fabrication, though. If truth be told, we do not know a great deal more about Columbus than about Leifur Eiríksson. Even if he never sailed to Iceland, there is more than adequate reason to believe that the existence of lands in the far west was not unknown in mainland Europe in the Middle Ages. As mentioned earlier, Iceland and Greenland were not as isolated in the heyday of the Vikings as they became later, when the North Atlantic climate deteriorated and trade declined. Guðríður Þorbjarnardóttir, pioneer of the Vínland settlement, made a pilgrimage to Rome, presumably around the mid-11th century. She must have told the story of her travels. Many other pious Icelanders made the journey to the Eternal City over the centuries, so knowledge of "Vínland the Good" could certainly have reached Rome quite promptly.

This hypothesis is supported by the fact that the first Bishop of Greenland, Eiríkur Gnúpsson, was appointed not only over Greenland, but *regionumque finitarum* (adjacent regions). This cannot be a reference to Iceland, which had its own bishops, so it appears to refer to other lands, farther to the west. Vínland is mentioned by that name as early as AD 1070, in Adam of Bremen's *Historia Hamburgensis Ecclesiae*, where he says that the King of Denmark has told him of a land of grapes and wild corn out in the ocean. Vínland is also mentioned in the annals of Oldericus Vitalis in the late 11th century. So we can safely conclude that the discovery of Vínland was soon known in Europe. But how long did that knowledge last? Could some hint of it have lasted four centuries? Could Columbus, growing up in the thriving port of Genoa, have heard some legend or myth which sparked his imagination?

In 1476, Columbus arrived in Lisbon, quite by chance. From Lisbon, his way to Iceland lay clear; merchant vessels from Bristol in the west of England plied each summer to Hafnarfjörður in Iceland, where they traded various sought-after commodities for stockfish (dried fish), which was Iceland's most important export at the time. In autumn and winter, the same ships sailed to Lisbon and points south, to sell Icelandic fish in exchange for exotic Mediterranean goods.

It was on one such vessel that the young Columbus is said to have travelled to Iceland. On the way north, the ship called at Galway on the west coast of Ireland, where Columbus heard of people with an Asian cast of countenance being washed ashore from a shipwreck. This strengthened his belief that Asia lay, quite accessibly, across the Atlantic.

When Columbus set sail in 1492 to look for Asia, under the patronage of the Spanish monarchs Ferdinand and Isabella, he did not follow the Vikings' island-hopping course across the North Atlantic, but simply set course for the west, and sailed on until he found land. Whether or not he knew of the Icelanders who had sailed in the direction of the setting sun five centuries before, Columbus was an explorer in his own right, following his instincts just as Eiríkur the Red and Leifur Eiríksson had done.

By an accident of history, Leifur Eiríksson's discovery of the New World changed nothing, while Columbus's rediscovery changed everything. In AD 1000, the nations of Europe were not yet casting greedy eyes upon the whole world. The Nordic mariners travelled, not as representatives of a nation, but as individuals. They only wanted a place to live and raise their livestock; warfare was not on their agenda, so when they met with forceful armed resistance from the indigenous inhabitants, they simply gave up the attempt and went home. And the Native Americans enjoyed another 500 years of isolation. When the Europeans made their second discovery of America, they came in force, under arms, and as conquerors.

Bibliographical Note

In my account of the Vínland voyages, I made use of the *Saga of Eírikur the Red* and the *Saga of Greenlanders* as published in the collected edition of *Íslendinga sögur* (Sagas of Icelanders) by Svart á hvítu (Reykjavík 1987). I also referred to these sagas and their accompanying notes in the *Íslenzk fornrit* edition, vol. 4 (Reykjavík 1935). In addition, I referred to *Landnámabók* and *Íslendingabók* in *Íslenzk fornrit* vol. 1 (Reykjavík 1968).

Ólafur Halldórsson's *Grænland í miðaldaritum* (Reykjavík 1978) provided useful information on the Nordic Greenlanders and Þorsteinn Vilhjálmsson's material on *Raunvísindi á miðöldum* in *Íslensk þjóðmenning* vol. 7 (Reykjavík 1990) was an invaluable source on navigational techniques and Star-Oddi. Both books have English summaries.

Details of the structure of Viking ships are drawn mostly from *The Viking Ships* by A.W. Brøgger and Haakon Shetelig (revised edition, Oslo 1971).

Archaeological excavations in 1961-8 at l'Anse aux Meadows are described in Anne Stine Ingstad's report *The Discovery of a Norse Settlement in America* (Oslo 1977). A more recent and updated report is *The l'Anse aux Meadows Site* by Birgitta Linderoth Wallace in Gwyn Jones's *The Norse Atlantic Saga* (second edition, Oxford 1986). In his article, *Was Vinland in Newfoundland?* in the *Proceedings of the Eighth Viking Congress*

(Odense University Press 1981), Einar Haugen makes several useful points.

The Norse Atlantic Saga by Gwyn Jones brilliantly recounts the story of Nordic exploration and settlement in Iceland, Greenland and Vínland, and draws attention to many of the problems of dating, interpretation and credibility. The book also includes translations into English of the relevant saga literature and a number of appendices confronting individual controversies and puzzles.

With regard to the possibility of a link between Columbus and the Viking voyages, I refer to Ólafur Egilsson's article, *Columbus in Iceland* in *Iceland Review*, 3/1991.

Further Reading:

Helge Ingstad, *Westward to Vinland*
Jonathan Cape, London 1969.

Gwyn Jones, *A History of the Vikings*
Oxford University Press, 2nd edition, 1984.

Gwyn Jones, *The Norse Atlantic Saga*
Oxford University Press, 2nd edition, 1986.

Magnus Magnusson and Hermann Palsson,
The Vinland Sagas
Penguin Books, Harmondsworth 1965.
Translations of the Saga of Greenlanders and the Saga of Eiríkur the Red, with an informative introduction.

Magnus Magnusson, *Vikings!*
Bodley Head, London 1980.

Magnus Magnusson, *Viking Expansion Westwards*
Bodley Head, London 1973.